Starlight Nights

Starlight Nights

The Adventures of a Star-Gazer

By Leslie C. Peltier

Foreword by David H. Levy

Illustrated with drawings by the author

Sky Publishing Corporation / Cambridge, Massachusetts

Dedication

To Dottie, who arrived on a comet

Acknowledgement

I wish to offer my heartfelt thanks to all those who, through their many acts of kindness over the past three score years, have helped to make my stars shine brighter.

In particular, to Edwin Way Teale, eminent naturalist, wise counselor, and steadfast friend, my debt of gratitude is quite beyond expression.

Published by Sky Publishing Corporation, 49 Bay State Road, Cambridge, MA 02138-1200 http: //www.skypub.com/

Library of Congress Cataloging-in-Publication Data

Peltier, Leslie C. 1900-1980
Starlight nights: the adventures of a star-gazer / by Leslie C. Peltier, illustrated with drawings by the author; foreword by David H. Levy.
p. cm.
Originally published: New York : Harper & Row, 1965.
Includes index.
ISBN 0-933346-94-8
1. Peltier, Leslie C., 1900-1980. 2. Astronomers–United States–Biography. I. Title.

BP36.P4 .A3 19999
520'.92
[B] 99-046982

Printed and bound in Canada

Contents

Foreword

By David H. Levy

NIGHTFALL IS SPECIAL FOR PEOPLE WHO LOVE THE SKY. WHETHER THEY ARE professional astronomers with state-of-the-art equipment or amateurs with small backyard telescopes, skywatchers everywhere anticipate the onset of night. At dusk on the evening of June 7, 1999, two dozen children sat on the floor of the observatory housing the 61-inch Kuiper telescope on Mt. Bigelow in Arizona. Around them, the walls and dome rose majestically to the sky. As the room darkened, I welcomed them not only as members of the University of Arizona Astronomy Camp, but also as privileged observers who had been awarded viewing time with this telescope in lieu of professional astronomers. The children were silent as I turned a small lever and the tall metal shutters began to part, inviting in the darkening sky and the light of the brightest stars. As the dome opened, I turned to page one of *Starlight Nights* and read about nightfall in a distant place and time:

> There is a chill in the autumn air as I walk down the path that leads along the brow of the hill, past the garden and the big lilac, to the clearing just beyond. Already, in the gathering dusk, a few of the stars are turning on their lights. Vega, the brightest one, now is dropping toward the west. Can it be that half a year has gone since I watched her April rising in the east? Low down in the southwest Antares blinks a red-eyed sad farewell to fall while just above the horizon in the far northeast Capella sends flickering beacon flashes through the low bank of smoke and haze that hangs above the town. Instinctively I turn and look back toward the southeast for Capella's co-riser. Yes, there it is, Fomalhaut, the Autumn Star, aloof from all the others, in a sky made darker by the rising purple shadow of the earth.

Reading these words took me back to a cloudy January 4, 1966, when I was in 11th grade. "I got permission this morning from Mrs. Lancey, my English teacher, to read *Starlight Nights!*" I wrote in my diary that evening. What a score for high school: I got to read an astronomy book and write about it for English class. I wasted no time getting started.

Many chapters later and well into the night, I was still unable to turn out the lights. To say the book captivated me is to profoundly understate its impact on my life. My comet-hunting career had begun 2½ weeks earlier, on December 17th. *Starlight Nights* would now serve as a road map for the journey. Thirty-five years and 21 comets later, it still does.

Peltier brings astronomy to life, from the telescopes he cherished . . .

> It seemed to me that if ever human attributes could be invested in a thing of metal, wood, and glass then this ancient instrument now in my keeping must long for one more chance to show what it could do.

. . . to his descriptions of comets and comet hunting:

> Time has not lessened the age-old allure of the comets. In some ways their mystery has only deepened with the years. At each return a comet brings with it the questions which were asked when it was here before, and as it rounds the Sun and backs away towards the long, slow night of its aphelion, it leaves behind those questions, still unanswered. . . .
>
> On the farm, in my early years of comet hunting, few sounds of civilization ever came to me inside my little dome. Even now, when I am hunting late at night, there is little to remind me of what century it is. In the dark silence of the dome two hundred years can disappear in just the twinkling of a thought and I am standing beside destitute, old Charles Messier as he knocks on the door of his friend Lalande and borrows a little oil for his midnight lamp. On another night I may, in spirit, be with Dr. Olbers of Bremen as he turns from stethoscope to telescope and begins his nightly search for asteroids and comets.

Many books explain *how* to observe the sky; *Starlight Nights* explains *why*. I have not encountered a single work that comes as close to capturing the passion of skywatching. Die-hard observers should lend the book to their families, as I did to my father. "I never fully understood or appreciated your passion for the night sky," Dad said, "until I read this book."

The Real Leslie Peltier

After I finished reading *Starlight Nights* back in the winter of 1966, the man who wrote it remained something of a mystery to me. The peace and satisfaction he described was so complete that I couldn't imagine anyone on Earth actually living so contentedly. Several months later I traveled to central Vermont for Deep-Sky Wonder Night, held in honor of *Sky & Telescope*'s popular columnist Walter Scott Houston. Not only did I get to meet the man we readers knew as Scotty, but I also was able to learn more about his close friend Leslie Peltier. Scotty told me that the voice of *Starlight Nights* was that of the real Leslie — a shy and retiring man who loved his Delphos, Ohio, home so much that he rarely left it. Scotty told how, in order to induce Leslie to attend the Spring 1932 meeting of the American Association of Variable Star Observers, he plotted with Leslie's sister Dorothy to kidnap the great observer and bring him to the meeting in Maryland. Scotty described how Dorothy had offered him the clothesline out back to tie Leslie up in order to get him into Scotty's car. It turned out to be unnecessary — accepting defeat, Leslie smiled softly, shook his head, and got into the car. I would love to have seen the delighted looks on the faces of those gathered at the meeting when the celebrated, but seldom glimpsed, Leslie Peltier walked in.

I resolved that night in Vermont that I would write to Leslie — not to the astronomical icon, but to the man. "I doubt he'll answer you," Scotty cautioned. "He rarely responds to letters." But two weeks later my spirits soared when I received an envelope with a Delphos, Ohio, postmark. Inside was a letter from Leslie, thanking me for my "kind words about *Starlight Nights*." He must have received many letters from would-be comet finders, for he was rather circumspect in his advice on the topic: "By far the greater proportion of comets," he wrote, "are now found on the photographic plate and this lessens, of course, the amateur's chance of making a find but I wish you the best of luck anyhow."

As my own commitment to comet hunting became more serious, Leslie's letters became more personal. "This has been a terrible winter thus far for any kind of observing and there seems to be no relief in sight," he wrote on January 8, 1968, in the midst of one of the coldest and snowiest winters I had known also, even from hundreds of miles away in Montreal. "The dome for my 12-inch is frozen tight under a foot of drifted snow and can't be moved. Today it is 10 below zero and will be colder tonight and you are probably having it equally as bad or worse. But if Winter comes can Spring be far behind?"

"You bet it can!" Dad quipped in reply.

Six years later, we were still exchanging letters. "Speaking of comets," he wrote on January 27, 1974, "Kohoutek has certainly been a disappointment. No doubt much will be learned from it but to the public it will cast some doubt as to the credibility of astronomers since they forecast it in such glowing terms. After Watergate and now Kohoutek just what can we believe?"

In April 1974, as the not-so-mighty Kohoutek receded from the Sun, I finally met in person the man whose life and work meant so much to me. Although we chatted about many things that evening, including Watergate, it was clear where Leslie's heart really lay when our talk turned to variable stars, comets, and clear nights. He showed me the old wooden tube that had once carried his 6-inch refractor, with the dates of each of his comet catches expertly carved into the mahogany. Looking at that tube was like looking at a Torah. The inscribed discovery dates jumped out, each one recording the triumph of a new comet find from years past, each one speaking to the future. As I sat in the Peltier living room that night, I realized that, just as Scotty Houston had said years earlier, the real Leslie was identical to the star of *Starlight Nights*.

The tube for Peltier's 6-inch refractor. The year of each of his comet discoveries is carved into the mahogany. Photograph by David H. Levy.

Peltier's Legacy

In 1977 I was busy writing my master's thesis for the English Department at Queen's University in Kingston, Ontario. Its title was *The Starlight Night: Hopkins and Astronomy*. Gerard Manley Hopkins, a famous 19th-century English poet and amateur astronomer, wrote wonderfully about the sky in his poem *The Starlight Night*:

> Look at the stars! Look, look up at the skies!
> O look at all the fire-folk sitting in the air!
> The bright boroughs, the circle-citadels there!

Could Leslie have known about this poem? I asked him in a letter. On October 28th he replied, "*Starlight Nights* was not named for the Hopkins poem and I am not familiar with it. It was not suggested by anyone or anything but just seemed a logical title and one that had never been used. Hope to have a new book out sometime in November. It's somewhat along the lines of S.N. but tells mostly of our doings here at our present home over the past 25 yrs. Only in its final chapter does it touch on astronomy and I'm not recommending it to *anyone*."

Leslie underestimated the beauty and grace of his new book. *The Place on Jennings Creek* is a hymn to the many other aspects of nature that inspired him, much as *Starlight Nights* is a hymn to the sky. It describes with loving exactness the natural surroundings of his home, from the flora and fauna without to the books within.

In August 1979 I stopped in Delphos for another visit with Leslie. I found him more subdued than he had been five years earlier. He was anxious about the future of *Starlight Nights*, the book he described as his "first love." Sky Publishing, which had contacted him about bringing out a softcover edition, had just published *The Cloudy Night Book,* and he was concerned that they might lose interest in *Starlight Nights*. "They published a cloudy night book," he remarked ruefully, "but I wrote the *clear* night book!" He needn't have worried — Sky came out with the book the following year.

As Leslie reflected that August evening on his long acquaintance with the night sky, I gained even more of a sense of what his tele-

Leslie Peltier and David Levy in front of the door to the Miami University observatory, Delphos, Ohio, August 1979. Photograph by Constantine Papacosmas.

scopes meant to him. He saw them not as tools, but as friends that enabled him to track the brightenings and fadings of variable stars and to discover comets. Even though he and I had come to know each other fairly well, I left him, as I had after our first meeting, with a sense of awe. I saw a calmness about him, a deep satisfaction with the relationship he had cultivated with the sky over so many years.

Less than a year later, in May 1980, Leslie died suddenly as he was preparing his telescopes for a spring night of observing. He never had his much-longed-for second view of Halley's Comet, and I never got to share with him the details of my first comet discovery on November 13, 1984 —59 years to the day after his own first comet find. But Leslie's list of accomplishments is immense, with 12 comets and 132,000 observations of variable stars.

A Message for the Future

On the first of March each year, Leslie had the habit of reobserving R Leonis, the star that, in 1918, ignited his love affair with variables. Three quarters of a century later, on March 23, 1993, I took two photographs of a region not far away, in Virgo. Carolyn Shoemaker found the faint images of a disrupted comet on those photographs. The fragments of that comet, Shoemaker-Levy 9, slammed into Jupiter in July 1994.

Those impacts vividly demonstrated that the comets that were so important to Leslie actually play an integral part in the evolution of our planet. We now know that ancient comet crashes were crucial events through which the building blocks of life may have been delivered to Earth. We also know that through impacts, the course of life on Earth has been severely altered, as when the dinosaurs were wiped out 65 million years ago.

"Time has not lessened the age-old allure of the comets," Leslie wrote in the last chapter of *Starlight Nights*. In ways he could hardly have imagined, the allure of comets has actually deepened with the years. As *Starlight Nights* appears again on the centennial of its author's birth, after having been out of print for just under a decade, it will inspire new generations like those attending Astronomy Camp. These young skywatchers belong to a technological age far removed from the rural peace of *Starlight Nights*, and yet Leslie's words call to them, inviting them to see beyond the sunset. With *Starlight Nights*, their generation — and ours — may confidently venture into the night.

Comets Discovered by Leslie C. Peltier

Number*	Name	Date
1925d	Wilk-Peltier	November 13, 1925
1930a	Peltier-Schwassmann-Wachmann	February 20, 1930
1932k	Peltier-Whipple	August 8, 1932
1933a	Peltier	February 16, 1933
1936a	Peltier	May 15, 1936
1937c	Wilk	February 27, 1937**
1939a	Kozik-Peltier	January 19, 1939
1943b	Daimaca	September 19, 1943***
1944a	van Gent-Peltier-Daimaca	December 17, 1943
1945f	Friend-Peltier	November 24, 1945
1952d	Peltier	June 20, 1952
1954d	Kresak-Peltier	June 29, 1954

* Comet designations are those that would have been familiar to Peltier.

** This comet was found by Leslie a few hours after Antoni Wilk of Cracow, Poland, discovered it. The comet was announced in Europe as Comet Wilk.

*** Leslie found this comet 16 days after its initial discovery by Daimaca of Bucharest, Romania, but at the time he was unaware of Daimaca's discovery because of war-related delays in communication. No other observer in the United States ever saw this comet, however, so his independent discovery was crucial in determining its orbit.

Pictures from the Peltier family albums

Clockwise from top left: Dottie Peltier, 1933; Dottie and Leslie on their honeymoon, camping in Texas at the edge of the Indian Spring; Dottie at the mouth of the Indian caves on the Rio Grande; camping in the southwest (*left*, their 1929 Ford and *right*, the 6-inch refractor, fitted with a metal tube for the journey); the Peltier farmhouse in Delphos, Ohio, 1929.

Photograph by David H. Levy

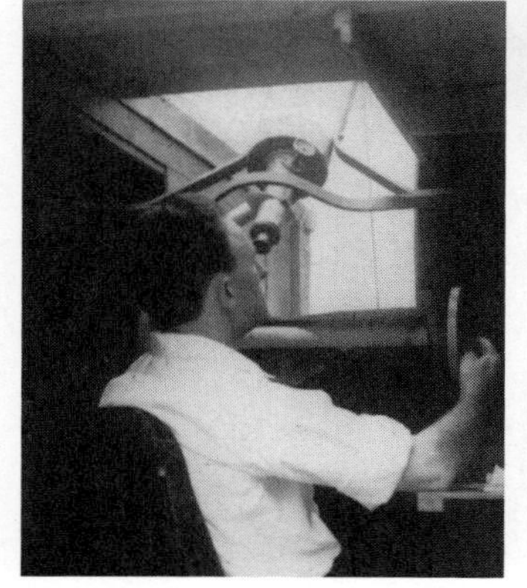

Clockwise from top: The cow-pasture observatory; the merry-go-round observatory under construction at the Peltier farm; Leslie seated at the 6-inch telescope in the merry-go-round observatory, 1979; Leslie at the controls of the merry-go-round observatory; the Miami University observatory, installed in the back-yard at Brookhaven, 1960; Leslie in front of the cow-pasture observatory, 1929.

1 A Starry Night

THERE IS A CHILL IN THE AUTUMN AIR AS I WALK DOWN THE PATH that leads along the brow of the hill, past the garden and the big lilac, to the clearing just beyond. Already, in the gathering dusk, a few of the stars are turning on their lights. Vega, the brightest one, now is dropping toward the west. Can it be that half a year has gone since I watched her April rising in the east? Low down in the southwest Antares blinks a red-eyed sad farewell to fall while just above the horizon in the far northeast Capella sends flickering beacon flashes through the low bank of smoke and haze that hangs above the town. Instinctively I turn and look back toward the southeast for Capella's co-riser. Yes, there it is, Fomalhaut, the Autumn Star, aloof from all the others, in a sky made darker by the rising purple shadow of the earth.

At the center of the little clearing the path ends abruptly for here, right at the top of the low hill, sit two stark-white structures. One of these is small and squat, no higher than my head, no wider than my outstretched arms. The other, standing boldly out against the sky is, by comparison, quite imposing. All day long they sit there side by side—these two—in sun, in rain, in snow, without a sign of life about them. It is only when the stars come out that they begin to stir. Then, like some snowy owl and owlet waking for a night of dark marauding, they spread apart their tight-closed wings, open wide their big round eyes and peer about in all directions. The prey they seek is hidden somewhere in the skies—for these are my observatories.

To the casual eye they would appear simply as a couple of oddly shaped buildings constructed of quite ordinary wood and metal, concrete and stucco. To me, these observatories and the telescopes housed within them are vital and alive, for they are compounded of the visual delights, the unexpected thrills, the lasting friendships, the expressions of good will and the multitude of kindred blessings that have come to me, all mixed with starlight, from the skies of three score years.

It is not yet dark enough to start the night's observing but I raise the windows and open wide the shutters of the dome to let the warm air trapped within escape. When I neglect to do this at the end of a sunny day the star images, which should be small and round and steady, will seethe and boil and frustrate until the cooler night air flows inside and makes them simmer down. Tonight, I notice that something more than just the daytime warmth is imprisoned in the dome. It also holds the daytime smell, the smell of fall, the smell of burning leaves.

To anyone who closely holds communion with the earth, each of the mild months of the year must have its own distinctive smell. To me April smells like freshly plowed ground; May recalls lilacs; the aroma of strawberries brings back long June days; July smells like new-mown hay; and other smells in season are musk-melon, chrysanthemum, and lastly, of course, October's pungent scent of burning leaves. These are all heady, potent smells and I close my eyes and inhale them deeply—with one exception. New-mown hay is one smell-of-the-month that I cannot abide.

Let Maud Muller rake her meadow. Let the lyric writers long for their fields of new-mown hay. I want no part of the hayfield, not even the smell! To me, hay has a hot heavy smell that brings back, all too vividly, hot, heavy work in the haymow. To this day I hold my breath when driving past a hayfield.

It now is eight o'clock, just dark enough at this time of year to start my prowl among the stars. Tonight the sky is clear, the stars are brilliant, and the definition, that all-important factor, should improve steadily hour after hour as the darkened earth gives up its store of accumulated heat. Already, even though some laggard light still lingers in the west, the southern Milky Way is flooding out around its murky midstream islands while here and there about the sky other kindling fires begin their silent clamoring for me to turn my telescopes on them.

I am fortunate in having two good telescopes at my disposal for this very purpose. One of these is a 12-inch refractor sixteen feet in length; the other, a mere midget by comparison, is a 6-inch instrument just four feet long. The 12-inch is thus about four times more powerful than the 6-inch for its lens has four times the surface area of the latter. However, each telescope has its own particular sphere of usefulness. Each one can perform its own specific duties much better than could the other one so there is really no cause for any rivalry between the two and I, for my part, have always done my best to insure domestic tranquillity by allotting them equal observing time.

I recall that in one of my old grade school readers there was a poem by Ralph Waldo Emerson entitled: "The Mountain and the Squirrel," which began—

> The Mountain and the Squirrel had a quarrel
> And the former called the latter, "little prig."

But the squirrel finally got in the last word—

> "If I cannot carry forests on my back,
> Neither can you crack a nut."

Sometimes, perhaps in the wee small hours of the night long after I have gone to bed, these two scopes may carry on just such a verbal battle with the 6-inch delivering the final squirrelish blow—"If I cannot see sixteenth-magnitude stars, Neither can you catch a comet."

On a night such as this, with its exceptional transparency, a special effort is always made to look for those objects which have eluded me on previous nights of only mediocre seeing. Tonight I glimpse an old and now long-quiescent nova at slightly below sixteenth magnitude and then I faintly glimpse a recently reported outburst of another star in a stellar universe far removed from ours. My next effort is even more successful as I watch a close pair of faint pulsating stars in Cassiopeia sparkling side by side as sharp and distinct as two tiny diamonds set against the velvet of the sky.

With the 6-inch I search for comets for nearly an hour low in the eastern sky where the late-rising moon will soon be coming up. It now is nearly midnight and so, as a final curtain to a gala spectacle, I let the scope glide slowly upward until, guided more by habit than by conscious help from me, it comes to rest on a misty little group of stars. Once again, as on uncounted other nights, I see:

> . . . the Pleiads, rising through the mellow shade,
> Glitter like a swarm of fireflies
> Tangled in a silver braid.
>
> —*from Locksley Hall by Alfred, Lord Tennyson*

So clear and sparkling is this autumn night that, with averted vision, I can see quite readily the wraithlike wisps of nebulosity that festoon and enmesh this entire little cluster. Something else I see too. Something wrapped in wisps of memory. Something that I always see each time I look at the Pleiades. I see a small Ohio farmhouse, a little boy, and a tall kitchen window that faced the east.

2 First Sightings

It was just after dark in the autumn of the year. As usual we—my father and mother, my older brother and sister, and I—were all in the dining room for here was the big Round Oak wood-burning stove, the center of our cool-weather universe. Here too was the large dining table which was always cleared after supper to make an arena for the evening's activities of school work or sewing or playing Flinch or Authors or dominoes. And frequently, as on this particular evening, this table also featured a most tasteful centerpiece—a large dishpan full of popcorn.

Popcorn, when properly salted and buttered, has one inseparable companion—water. And at this particular moment I just had to have a drink! Sometimes, in anticipation of these frequent yearnings, a pitcher of water and some glasses would also grace the table but on this occasion these accessories had not been provided so I went into the kitchen to get my drink. Now when one did this at night in a 1905 farmhouse one did not pause at the kitchen door and snap on a light switch. There was no light switch, for electricity was still twenty dark-years from the farm. So I just left the door wide open and, aided somewhat by the feeble glow that managed to struggle through from the shadeless kerosene lamp in the dining room, I groped around until I located the water. Farmhouse kitchens in the year 1905 were also completely uncluttered with any such folderol as sinks and faucets. At our house the last chore of the day was for Dad to bring in a large bucket of water and set it up on the kitchen table.

Quite often, however, this table was already fully occupied and the bucket of water got no farther than the floor and sometimes on these occasions, when we all were busy around the table in the dining room, through the open kitchen door would come an unmistakably liquid, "Tlot-tlot, in the echoing night." Then would come the measured tread of toenails on the kitchen linoleum and our black terrier, Buster, would stroll back in with a telltale drop of water still clinging to his graying chin, give us a look filled with gratitude for our thoughtfulness, then stretch out back of the stove and return to his panting, paw-twitching dreams of chasing rabbits.

On this night the water bucket was on the table but it had been placed somewhat beyond my modest five-year reach. Just a few months before I had encountered this same situation and that night I had assayed to climb right up on the table in my quest for water. Unfortunately, I had climbed up on the opened table leaf thereby promptly upsetting the entire table. I got the water, all right, along with a crashing avalanche of pots and pans. All this was still quite fresh in my memory so this time I sent out an appeal for aid. My mother, who also remembered, came to my assistance at once. She got me the glass of water and, as I stood there drinking it in the shadowy dark of the room, I could see the stars shining brightly through the east window before me. About halfway up the sky I noticed a little group of stars and pointing to it I said to Mother, "What's that?" Her eye followed along my outstretched arm; "Oh," she replied, "Those are the Seven Sisters, sometimes they are called the Pleiades." This was my first meeting with the stars.

Two years later when I was seven my orbit once again crossed that of a star, or, to be more exact, a planet. On this occasion we had all been away from home in the afternoon and it was quite dark when we returned. On a farm there were still the regular chores to do so I went along with Dad and carried the lantern while he fed and watered the stock. We had just finished feeding the pigs and were about to start back to the house when Dad suddenly stopped and pointed high up in the southeastern sky. "Look up there, son," he said, "there's another lantern—that's Jupiter."

These two early encounters with the things in the sky must have left me somewhat star-struck for I still remember how thrilled I was when I went to school that September to find, on the last page of the Second Reader, the picture of a sweetly smiling moon and beneath it these lines:

> "Lady Moon, Lady Moon, where are you roving?"
> "Over the sea, over the sea."
> "Lady Moon, Lady Moon, whom are you loving?"
> "All that love me, all that love me."

It was the beginning of a long and happy friendship.

Sometime during these years, I can not recall just when, I made the acquaintance of the Big Dipper. This is by far the best known of all the star figures and must surely be familiar to nearly everyone who has ever lived in the country. For many of the farmers in our region the Big Dipper served as a weather indicator. As a child, even before I knew this constellation, I can remember being intrigued and mystified whenever I would hear such remarks as, "It'll surely rain today, the Big Dipper was turned upside down last night." Or, "Watch out for bad weather, the Dipper bowl is turning over." Strange remarks indeed from those who live so close to nature as the farmer. How few there are who have ever watched the circling of the stars about the pole.

To the farmer the weather is a vital thing and here on the farm we had many weather signs and portents and most of them, unlike the Dipper's doings, were logical and reliable. I had early learned the familiar rhyme:

> Red sky in the morning,
> Sailors take warning.
> Red sky at night,
> Is the sailor's delight.

I found that it was seldom wrong. I also knew that a heavy morning dew and fleecy flat-based clouds were both signs of fair weather and that whenever the telephone wires hummed louder than usual or when the train whistle half a mile away sounded extra clear then we could expect either rain or snow. My brother and I had a weather sign of our own that no one else seemed to know about. We shared a small upstairs bedroom during all our youthful years and sometimes at night, after we had blown out the lamp and gone to bed, we would notice that suddenly all the roosters in the surrounding country had started to crow. Their clamor would continue for perhaps ten minutes and then gradually die out. Invariably, an abrupt change in the weather would take place before morning.

I have no recollections of any significant sky happenings during my eighth and ninth years, other than those that took place in the skies of earth, for it was in this period that I followed with intense interest the fickle fortunes of the rival airmen, Hubert Latham and Louis Bleriot, in their efforts to fly the English Channel. It was shortly after this milestone in aviation history had been achieved, in 1909, that my cup of joy and anticipation quite ran over when I learned that an airplane—a real airplane—was coming to the nearby town of Delphos and would make a flight from there on Labor Day. This was only three weeks off and, in spite of the fact that school would begin on the day after, I could hardly wait until that Labor Day arrived. This was, I am sure, the one and only time that I ever displayed any such eagerness for the end of vacation.

The Great Day finally came and with it came perfect weather. Shortly after noon a freight train pulled into the station and there, on a flatcar, looking, I thought, somewhat humiliated by such down-to-earth transportation, was the airplane, the first that I had ever seen. The plane was rolled down a ramp to the ground and from there it was picked up bodily by about twenty husky volunteers and carried about eight blocks down Main Street to the city park where a makeshift runway had been roped off in an adjacent field. Here the airplane carriers set down their burden.

A crowd of the curious immediately began to form around the plane for this was a first for nearly all of us. Gradually, I wormed my way through the milling wall of spectators and finally emerged, as I had hoped, near one wing tip. Here I had, quite appropriately, a good worm's-eye view of most of the plane. Fascinated, I feasted my eyes on every detail of the plane that I could see; the light wood framework, the cloth wing covering, the crisscross maze of piano-wire braces, the tail and rudder, the three bicycle landing wheels, and finally the single pusher-type propeller. Then with a sudden daring thought, I stole a look at the crowd around the plane. No one seemed to be watching me. I looked for the pilot; he was busy adjusting something on the opposite side. I edged just a little closer then quickly reached out and up until my hand made contact with the wing. A dream had just come true. I had touched—actually touched—an airplane!

When everything was finally in readiness for the take-off we all were shooed back to a respectful distance and the pilot, a young man named Bud Mars, started up the engine after a few pulls on the propeller and mounted to the flat seat just in front of it. Here he put on his black leather gauntlets, reversed his cap, pulled down his goggles, and waved to the cheering crowd. With a roar the engine blasted back its small cyclone, there was a forced take-off of all the straw hats behind the plane and then the machine started rolling down the runway. After a spurt of not more than two hundred yards the wheels left the ground and the crowd gave out a lusty cheer. The flight proved to be completely without incident. At no time was the plane more than three hundred feet above the ground and the entire distance covered could not have been more than two or three miles. But everyone seemed more than delighted with our air-age premiere. I know I was.

Too much credit can never be given to these early bird-men. They were pioneers in the truest sense of the word. They gambled everything on the uncertain performance of these crude machines—and frequently they lost. Many of them—including Bud Mars—were to die with their goggles on.

Actually, when I first watched from my grass-top viewpoint, the plane, a Curtiss biplane, had made me think of a two-story Chinese pagoda. It had none of the graceful lines of the Antoinette and Demoiselle monoplanes whose flights I had been following. But nevertheless, to me that day, it was a thing of beauty. By the time we reached home that evening my mind was fully made up; my future career was now a certainty. I too would be an aviator.

In less than a week I was in the air. I could look down over the edge of my wing and there was nothing but empty space and the dark earth down below. I could look to the right or to the left and see the woods, the cultivated fields, the farmhouses in the distance all around me. I could turn my rudder and make a complete tight circle. I could push forward my elevator lever and go into a steep nose dive.

What did it matter that I had no ceiling—that a thick canopy of oak leaves closed in my entire sky—I could always fly blind. What did it matter that I could only enter my plane by crawling out on the big oak limb and sliding down the rope by which the plane was suspended. Some day I would fix all that with a block and tackle. What did it matter, in fact, that a single sheet of corrugated roofing made the wing of my plane, that a wide plank furnished a fuselage and a pilot's seat, and that an oil drum lid became a rudder? No plane was ever built which served its purpose better. My flying time logged only happy hours.

All over the world this had been a tranquil era. There had been peace on earth and also in the heavens. But those placid skies did not prevail for long. Already something had been sighted. Strange and awesome nights lay just ahead.

3 The Comet Year

IN 1910, ALL OVER THE WORLD, CURIOUS AND OFTEN FEARFUL EYES were turned to the heavens. There was much to see and ponder. In the middle of January a comet, known simply as 1910a, unannounced by any trumpets of the sky, came suddenly from behind the sun. On January 18 astronomers at Lick Observatory in California saw it at midday just east of the sun and estimated its brightness as exceeding even that of the planet Venus, which was only a short distance away. Here on the farm, by the time we learned of the comet from our twice-a-week newspaper it had moved much to the east and north and its light was considerably diminished, though it still was quite conspicuous in the western sky and our entire family watched it for many nights. I still have a vivid mental picture of this comet just as I saw it then, through the leafless branches of the young walnut trees near our house. The trees have long since grown to man's estate but the image has not aged. From it and a host of other images of later comets my mind's eye can estimate with confidence that when we saw it then, the object was still of first magnitude and I can still see that long feathery tail that curved upward to the east and south.

In almost any other year Comet 1910a would have been justly acclaimed as a magnificent comet, but in that spring of 1910 the world was anxiously awaiting the arrival of Halley's Comet, the most famous comet of them all and the first one definitely known to be periodic. Previous to the appearance of a bright comet in the year 1682 all comets were believed to be strangers from the depths of space who, in passing, paid the sun a brief visit and then departed never to return. The man who completely unmasked the visitor of 1682 and forced it, and a host of other comets as well, to reveal their solar family ties was the famous English astronomer, Edmund Halley, one of the great observers of all time and a genius of the highest order.

By his accurate observations of the comet that year Halley was able to show that it was traveling in an elliptical orbit. Since an ellipse is a closed orbit this meant that eventually the comet would return and that, like the planets, which also travel in elliptical orbits, it too was a captive of the sun. He determined the period of the comet to be about seventy-six years and predicted that it would return in 1758 or 1759.

During the course of all this investigation Halley noted that similar bright comets had previously appeared in 1607 and 1531. As all of these dates were separated by the same interval of seventy-six years he announced that they all were apparitions of the same comet but that final proof would have to wait until the predicted next return.

On December 25, 1758 the comet made its reappearance, though Edmund Halley was not there to see his Christmas present to the world. He had died sixteen years before at the age of eighty-five, after a long and useful life of service to astronomy and to humanity as well.

Two trips later we watched the comet from the farm. At this latest return it had been sighted by large telescopes as early as September 1909 while it was still far, far out in space and during the months that followed it had been approaching the sun with ever-increasing speed.

Here on earth the majority of those who knew of the coming event awaited its arrival with eager anticipation. But there were others, the timid and the superstitious, who followed the nightly advance of the comet with dread, refusing to accept the common knowledge that the visitation would be but a repetition of what was known to have occurred at seventy-six-year intervals ever since the object was first recorded in 240 B.C. Suicides were said to have been numerous that spring. Many persons disposed of their property in anticipation of the end of the world and the hell-fire preachers made the most of a visible sign in the heavens.

It was truly a noble comet and, though seen best from the earth's Southern Hemisphere, even here we saw it well, both in the east before sunrise and, a few nights later, in the western evening sky. For us it was a more spectacular comet than 1910a had been but, nevertheless, I still have a more vivid mental image of the earlier comet because of the greater impact of a first impression. They both are cherished memories.

The orbit of Halley's Comet is a long narrow ellipse which rounds the sun well inside the orbit of the earth, while the opposite extreme lies somewhat outside the orbit of Neptune. The most amazing thing about an orbit of this type is the extreme variation in the speed of the comet that travels it. In 1910 Halley's Comet circled the sun at a speed of about thirty miles per second, while in 1948, at aphelion—or its greatest distance from the sun—it had virtually come to a standstill. Then slowly, almost imperceptibly, it began the long progressive speed-up that finally will culminate at its next reunion with the sun in 1986. Nearly half of its period of seventy-six years is spent in making the cold, dark, outer loop that extends beyond the orbit of Neptune.

As this is written the comet is still moving along this outer loop that it began to make in 1930. In 1966 it will recross Neptune's orbit now headed toward the sun and with only twenty years remaining before its scheduled arrival. The sun, if viewed from the tremendous distance of the aphelion, would appear simply as an extremely bright star without any perceptible disk. The comet itself would, in all probability, be a dark, invisible body made up of a more-or-less porous aggregate of smaller particles.

It is presumed that in the case of a comet such as Halley's, with a long, narrow, elliptical orbit, nothing but the nucleus of the comet ever makes the complete round trip. The other parts, the tail and the coma—which is the nebulous envelope surrounding the nucleus—are manufactured from the nucleus only as the comet nears the sun. The amount of solid material in the nucleus must either be small or widely scattered for comets have been observed to pass directly over stars without appreciably dimming the star's brightness. On May 18, 1910 Halley's Comet crossed directly between the earth and the sun. An expedition had been sent to Hawaii to observe this transit with a 6-inch refractor but could find not the slightest trace of the comet's passage. This would seem to indicate that the nucleus is of a porous nature for a solid body no larger than twenty-five miles in diameter would have easily been seen as a black speck moving across the face of the sun.

After 1966 the comet will spend the next decade in traversing the vast void between the orbits of Neptune and Uranus. A few years later, when it is somewhere between the orbits of Saturn and Jupiter, the emulsioned eyes of the great reflectors, both on earth and in orbit, will hunt the comet down and within the following few weeks its future nightly course will be precisely plotted.

Within historic times twenty-eight visits by Halley's Comet

have been recorded. On an early trip it witnessed the defeat of Attila's Huns in A.D. 451. It arrived in time to preside over the Norman Conquest in 1066. In the year 1456 the menacing appearance of the comet so alarmed Pope Calixtus that he decreed several days of prayer and established the midday angelus. With a great clanging of bells he then besought the comet to visit its wrath solely on the invading Turks. In 1607 it was watched by both Shakespeare and Kepler and I like to think that it was also seen by Captain John Smith and Pocahontas in the frontier skies of Jamestown. On its following trip around in 1682 the comet was observed by Halley himself, who probed into its periodic past and bequeathed to it an honored name that it can bear with pride throughout the solar system. By 1835, when it returned, affairs of earth had speeded up. Many a canal boat traveler, looking down, could see the comet glowing on the surface of his highway. Man himself had taken to the skies when the comet last appeared in 1910, for he was making fledgling flights of perhaps one hundred miles. In 1986 our historic visitor will be visited in turn, for in that year a spacecraft from the earth will hold a rendezvous with Halley's Comet out in space!

Who would venture to foretell the wonders and achievements which the comet will witness in that distant year of 2062? Or will man himself prove periodic? Will the Huns be back again?

Inasmuch as Halley's Comet is such an infrequent visitor to our celestial shores it is no small attainment for a person to observe and clearly recall two appearances of this noted voyager. Only an octogenarian could hope to qualify for such an honor and even he (or she) would have to be very discriminating in their selection of the proper year in which to be born, in order that they be old enough to remember the first apparition and then not too old for the second.

On one of those May mornings in 1910 we had watched the great tail of the comet shimmering in the eastern sky just before dawn. So swiftly did the comet dash around the sun that when next we saw it two nights later that tail was glowing in the western evening sky. We later learned that during that intervening night, while we all slept, the earth had passed right through the outer reaches of the comet's tail.

Four miles away on that same night, while all the earth was wrapped in this frail filament, another event had taken place. But of this, until later, I shall not tell, for its full import I was not to know until a score of years had come and gone and another star attraction was shining in my skies.

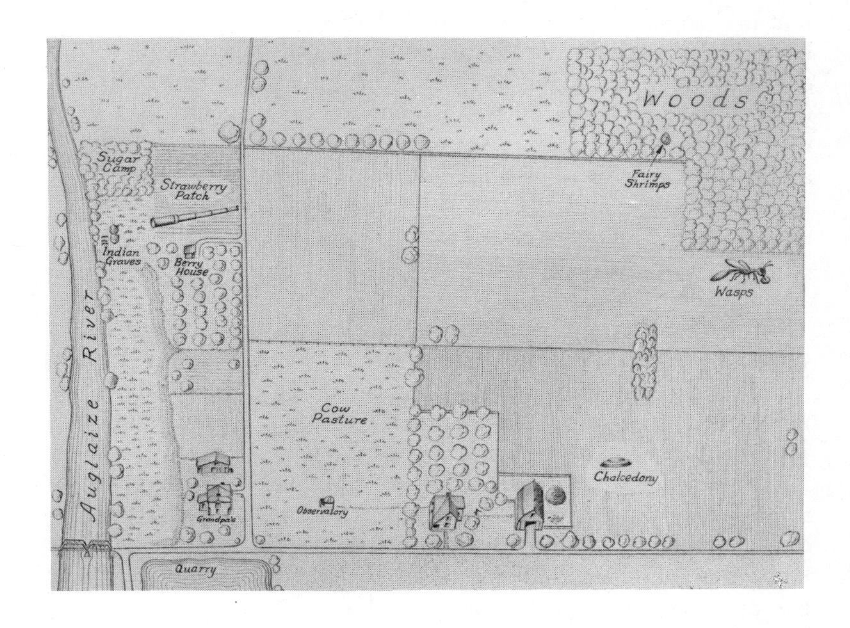

4 The Farm

WERE I TO WRITE MY OWN SET OF BEATITUDES I WOULD PLACE NEAR the top of the list: "Blessed are they who are raised on a farm." Nowhere else in the whole wide world can one acquire a keener appreciation for the things in life that are really worthwhile. Nowhere else can one achieve a deeper sense of freedom and independence than on a farm, and I seriously doubt that anyone who has been brought up on a farm and who loves the farm and farming can ever be completely happy in any other environment.

Our farm is located in northwestern Ohio, about twenty-five miles from the Indiana line. It lies in a region known in pioneer days as the Black Swamp. A region where malaria was so prevalent and so severe that it is said that this area could never have been settled without the aid of quinine. Local histories state that even the dogs belonging to the settlers constantly suffered from fever and ague. Now, all this is changed. The swamp has been drained; the trees and dense undergrowth have been cleared away. It now is a land of small and fertile and well-kept farms. It is a land of corn and hogs, of cows and clover. A land of wheat, soy beans and sugar beets; of Granges, Ladies Aids and country churches. Also, of late, it is a land of price supports, allotments, and controls.

For generations my ancestors had been farmers. On my father's side they were among the first settlers in this region and I have, as a prized bit of proof of this, the original land grant, signed by President Polk and dated February 1, 1849, giving title to a portion of this area to my great-grandfather.

We had about fifty acres under cultivation—most of it in rotational crops of corn, followed by wheat or oats, and then by clover. The corn made its way to market in the form of fattened hogs, and the clover either as seed or turned into cream by our five or six cows. Until the tractor took over in the early twenties we always had three or four draft horses and hay and shredded fodder filled the barn's huge mow. The farm fronted on a good country road, now black-topped, but, even in earlier days, always well maintained. Half a mile to the south is the Pennsylvania Railroad and right beside it—until the late teens when it conceded to the automobile—was an electric interurban line where, for a fare of ten cents (five cents for children), we could ride the four miles west to the town of Delphos, or for fifteen cents we could make the twelve mile trip east to the county seat at Lima.

Our house was a simple and practical two-story six room structure resting on a well-masoned foundation of flat stones which had been quarried from the nearby river bed. Both the house and the red-painted barn were well located among large elm and maple trees and there was a single linden tree just behind the house which, each blooming season, was loud with the humming of thousands of honeybees working on its pendant flower clusters. As children, we evoked other, shriller sounds from this same tree by making wooden whistles from its branches. We would select a branch about half an inch in diameter and cut off a section about three inches long. We would then gently pound this section all around with the handle of the knife until the bark was entirely loose and could be slipped off the wood core. This core was then cut to the proper whistle shape and then the bark cover was pulled back on. Dad told us that it was from the polished smoothness of this wood core that we got the expression, "as slick as a whistle."

Along the west side of the yard was a row of walnut trees that Grandpa had planted long ago by simply digging a hole for each one and dropping in a walnut. He once told me that he had to do this operation twice for just as he had finished planting the last hole he discovered that a couple of his hogs had been following him, rooting out and eating every walnut. Between the house and barn we had a thriving young orchard of apple, peach, and cherry trees, while back of the house was a fenced-in chicken yard. My mother's quite extensive rock garden lay along the western border of the yard beneath the walnut trees and the remainder of this grassy front area was devoted to our various games and athletic endeavors. One summer day this front yard also became, for me, a disaster area.

Nearly all of our daily traffic lay between the house and barn and here we always had a good all-weather cinder path, but no one had ever bothered to build a walk of any kind out to the road and the mailbox. This oversight was always quite annoying in wet weather so one day I decided that I would do something about it. After much careful thought I laid my plans for a walk of irregularly-shaped stepping stones, to be made of cement molded right in their permanent location. I dug out a strip of ground about two inches deep and the width of my walk-to-be. I then built the low forms, spacing them all about two inches apart so that I could fill these spaces with dirt planted with grass seed and eventually have a narrow strip of grass between each stone. With our steel-tray wheelbarrow and a long-handled shovel I started out to mix up the first batch of cement. My first stop was at the barn where we usually kept some sacks of cement in a corner of the storeroom. Next, at the sand pile outside, I added the proper proportion of sand. Then on to the well near the house where the water was poured in with the aggregate and thoroughly mixed by hand. Finally the soupy mixture was wheeled on to the front yard and shoveled into the forms until they were filled to the brim. This same ritual had to be repeated many times and I finished just in time for supper.

Next morning I went out to remove the forms and plant my grass seed between the stones. When I knocked off the first form I noticed that the edges of the stone crumbled badly. I poked the stone with my finger; it was the same mushy mixture it had been the night before! Completely puzzled, I reviewed in my mind each step of the mixing process. There seemed only one possibility; I might not have allowed sufficient time for the cement to set. With a sudden thought I dashed out to the barn. Perhaps the cement sacks had the setting time imprinted on them somewhere. Over in the dark, windowless corner of the storeroom I picked up one of the empty sacks and carried it out to the light. I found not a word about setting time but I did find, in big, bold, sickening letters, the words ARMOUR'S FERTILIZER.

About two hundred yards to the west from our house the road spanned the leisurely Auglaize River over a high, plank-floored iron bridge. Somewhere on the opposite side, no one knows precisely where, the present road crosses a now obliterated path that roughly followed the winding course of the river. This was one of history's highways. Doubtless first marked out by the herds of buffalo that long ago roamed this region, it later became a well-traveled trail for successive tribes of Wyandotte, Miami, and Shawnee Indians. This trail has known the tread of Little Turtle and Tecumseh and the Prophet. Here, passing in pursuit of renegade or Indian, have gone the scouts, Lew Wetzel and the

Zanes. Finally, in 1794, came General Anthony Wayne who, with his army, built a string of forts along the river and soon brought lasting peace to all this border region.

On the north side of the road, just before it crosses the river, lies a long, narrow strip of river bottom land that follows along the stream for about a quarter of a mile. This strip of wasteland was a part of Grandpa's farm and the large farmhouse where my father, his three brothers, and two sisters all were born and raised, still sits on a commanding site at the crest of a long hill near the road. It was along this winding plot of pasture and woodland that I spent many summer hours of my early boyhood for during all these years I was our family's keeper of the cows.

On a level, highly productive grain farm it does not pay to devote much acreage to pasture for livestock, especially when, as in our case, there was plenty of free grass down along the river. So, each morning right after milking time, it was my pleasant chore to escort our little herd of half a dozen cows down the dusty road to their river pasture. On arriving there the cows always went directly to the river, then, after standing knee-deep in the shallow water for a long drink, they gradually scattered out to whichever part of the pasture most appealed to their fancy. They usually ate quite contentedly for a couple of hours which left me free to read a book or hunt English sparrows with my air rifle or to just wander around and engage in a great miscellany of early endeavors.

It was during these bovine breakfast hours of my early years that I learned to swim; caught a six pound carp; stayed under water for a full minute; captured a sleeping snapping turtle by the tail; ate too many green apples; memorized Thanatopsis; and played my fife.

This fife had loomed large in my boyhood days. I had "sent away" to a mail order house for it at a total cost, I well remember, of just twenty-three cents—including postage. It was a simple tin affair that could be played by anyone with a pair of lungs and a total of six fingers and when Grandpa heard me playing it to the cows down along the river I had to order him one just like it. Grandpa had soldiered under Sherman through the Civil War and he took to fifing like a duck to water. The fife was great company for him and, no doubt, it brought back a lot of martial memories. For many years, on balmy evenings when the wind was right, the strains of "Marching through Georgia" and "Tenting on the Old Camp Ground" would come to us across the field from the darkened porch of the big house where he lived all alone.

About midway in my long pasture paradise was a high knoll which here formed the east bank of the river. Innumerable springtime floods had cut the bank at this point to make a steep cliff about fifteen feet high. Here, at the summit of this knoll, was my lookout. From this lofty vantage point I could survey the winding river banks in both directions and if I happened to be watching at just the right time I could usually forestall the all-too-frequent attempts of my charges to wade across the river to sample the neighbor's succulent green corn in the field on the other side.

This idyllic little knoll had a most intriguing name. It had always been known as "The Indian Graves." About ten feet back from the edge of the bluff, surrounded by tall oak trees, were three graves. Two of these were side by side, the third lay about six feet to the north. Although they must have been very old the outlines of all three graves were still plainly visible and each spring the low mounds were covered with a carpet of deer-tongues and spring beauties. Without a doubt they were Indian graves and somehow, from my early childhood, I had nurtured the understanding that here on this little hilltop were buried some of my own early ancestors.

Whether there is any factual basis for this belief I do not know. It could have stemmed from something told me as a child, or it might simply have been a wishful thought of my own. I will never be certain. I only know that it was well within the realm of possibility for on both sides of my family some of the earlier stalwarts had become quite involved in Indian affairs. Early one morning in the year 1812 my mother's great-grandfather was shot and killed in a surprise Indian raid on his little log cabin in central Ohio. His family was rescued by soldiers from a nearby fort. On my father's side, in contrast to this carnage, one of his remote forefathers went to quite the opposite extreme and married an Indian girl, and from somewhere along this ancient branch of my ancestral tree, I liked to believe, came those three fallen leaves that rested on my hilltop.

5 Early Days

STANDING OUT CLEAR AND SHARP IN THE MISTY DAWN OF MY earliest recollections is a ladder. This was a ladder about twelve feet long and quite ordinary in every respect save for its location—it was indoors. It was our only means of reaching the three bedrooms on the second floor of our still unfinished farmhouse. To an active three-year-old this ladder was far more fun to climb than any stairs could ever be.

All too soon Dad got the stairway built and the ladder went back outside to lean against the cherry tree. The new stairway was a thing of beauty and it had some features that helped to make up for the loss of a ladder. It had wide treads and low risers and in winter those steps made the warmest seats in the house for the stove was just below. It had fancy turned spindles and a smooth hand rail that provided a fast but all-too-brief slide that ended with impressive finality at a very solid newel post.

As soon as the stairway was usable the plasterer came and plastered the walls and ceilings of the bedrooms. One day I heard Dad address the white-overalled mason as "Mr. Chambers." I reached up and gave a tug at Dad's sleeve, then, pointing to the old gentleman wielding the trowel, I looked up at Dad and whispered, "Mr. Pots."

Dad had designed and built our house. He was quite adept at carpentry or, in fact, at anything requiring manual dexterity. Along with this he had an abundance of artistic talent for a number of his oil paintings hung on our walls. His was a well-balanced make-up for he combined both the practical and the aesthetic. He had an equally keen eye for tall corn, a well-grown porker, a Harrison Fisher cover on the *Saturday Evening Post,* or a game of billiards at the Masonic Hall. For years he was master of both the lodge and the local grange, he never missed a Farmer's Institute, and he was the first in all this region to try out that newfangled crop—soy beans.

Dad was six feet tall, slender, and erect. He had clean-cut features and his coal-black hair was still dark and heavy at seventy-five. In his younger days he sported a neat mustache and was said to have cut quite a dashing figure at the handle bars of his shiny high-wheeled Columbia bicycle.

My mother was of medium stature, a little on the plump side, and had light auburn hair. She had taught school for several years prior to her marriage and was exceptionally well versed in history and American literature. She could quote from the New England poets and from the Bible on any and all occasions. Even after twenty years' absence from the classroom she still remembered all the important dates of history and all the capital cities of the world. One of her most outstanding traits was her unfailing optimism. She always saw the brighter side of everything and everybody. In her garden it was always the flowers that she saw—not the weeds nor the aphids. "In the long run," she often said, "things usually happen for the best."

Mother and Dad always seemed to be in complete harmony. They even had pet nicknames for one another which they frequently used around home. She was "Judy" and he was "Deacon." I know of no reason for the Judy for her name was Resa, but I am sure that the Deacon earned his title by having served in that capacity either in church or in lodge—probably both.

Our house on the farm had no plumbing, no running water, no central heating, and no electricity but, as neither had any of the other farmhouses, we all were quite contented. Like most farmers we always had plenty to eat and I think that it was quantity that we appreciated much more than variety. In our early years, particularly, our suppers sometimes would consist simply of mush and milk, but I do not recall that we ever complained of this fare for it was good and it was filling. We butchered twice each winter and throughout the year we had all the milk, butter, eggs, and cottage cheese that we could possibly use. We raised our own fruit and vegetables and, of course, canned most of the things we needed to tide us over the winter months.

About a mile from our farm was the large country church which also served as the cultural and social center for the community of several square miles of farms. We all attended every Sunday. Both Mother and Dad were teachers in the Sunday school and were extremely active in all the affairs of the church with one exception—the revival meetings. Every winter these rousing get-togethers were held nightly for two long weeks as a sort of faith tonic for the faltering; a booster shot for backsliders. Some of those old-time preachers could work up quite a storm and I always felt that the success of each night's meeting was judged on a sound basis. Mother and Dad usually attended these revivals but they watched them from a neutral corner. In fact, Dad, on more than one occasion, had voiced his opposition to this whole business of seasonal religion. He always noted that the fieriest testifiers were the ones who cooled off the quickest when the meetings finally came to an end.

It was during this same renascent era that, out of curiosity, we drove fifteen miles in our family carriage one night to see and hear the famed evangelist, Billy Sunday. At the height of his career as a baseball player Billy had forsaken the National League outfield in favor of the sawdust trail. But he took all of his big league energy right along with him and he made those meetings something to be long remembered. Only a sound camera—still uninvented then—could have properly preserved the drama of Billy's violent assaults against the citadels of sin.

Mother was a life member of the Woman's Christian Temperance Union and proudly wore her little white satin bow on all occasions. She also was a zealous leader in the affairs of the Woman's Foreign Missionary Society and was deeply concerned with the welfare of the heathen—particularly those in China, Korea, and India. This interest was, in a way, passed on to me on one occasion, for it was on a Sunday morning, amidst surroundings of missionary motif, that I made my debut as a public speaker. On this particular morning a returned missionary from India had addressed the Sunday school congregation. He had talked at length about foreign missions in his field and had gone into considerable detail about the caste system then prevailing throughout India. His lecture was followed by an informal discussion between the speaker and various interested members of the congregation. It now was nearly noon so, at the first brief lull in the questioning, the superintendent arose and, in an attempt to terminate the affair, asked in a negative tone if there was anyone who had anything further to say on the subject. Indeed there was; I had something!

My mother, some time before, had read me a little poem which I thought quite funny and I had memorized it. This seemed to me the perfect time to deliver it. I was about six years old and, along with a number of contemporaries, was sitting in the front row of seats. I quickly mounted the two steps to the rostrum, turned about, made my little bow and gravely recited

The poor benighted Hindu
He does the best he kin do.

He sticks to his caste, from first to last,
And for pants he makes his skin do.

It was a most appreciative audience and I got a big hand.

Considering the fact that we spent our youth in a world without radio, television, movies, record players, and automobiles a youngster of today may well wonder how we passed our time and what we did for entertainment before these modern marvels came along. Actually, I am sure that our days were just as filled with living as are any present days. But they were not supervised days; they were not organized days. We were left to our own devices and we provided our own entertainment.

We had games, we had pets, we had projects and we had a multitude of hobbies. Best of all we had lots of books, for we all loved to read and we subscribed to a good many magazines. Each Sunday, at church, we were given the two juvenile magazines, *The Classmate* and *The Advocate.* Then, on the way home we would exchange our *Saturday Evening Post* for a neighbor's *Youth's Companion.* Here at home we took *The Farm Journal, Green's Fruit Grower, Farm and Fireside,* and *The Ladies Home Journal.* Of this last I can only recall the one full page which each month showed the doings of the Brownies, by artist Palmer Cox.

For years we took a family magazine named, or perhaps I should say, misnamed *Success.* Dad thought he saw a long-range bargain here and took out a life subscription on this for which he paid about twenty-five dollars. A year or so later *Success* was a failure and stopped publication. Still another of our magazines took the final count in this era. Dad had always been an ardent admirer of the great Teddy Roosevelt and consequently the political news magazine, *The Outlook,* which was published in the interests of Teddy's Progressive party, was a regular visitor to our house until the elections of 1916 when both the Bull Moose and *The Outlook* dimmed and went out.

In our living room at the farm were two bookcases filled with books. One of these, a homemade affair beneath a window, held a twenty-five volume set of an early edition of the *Encyclopaedia*

Britannica. The other was a high, golden-oak combination desk and bookcase that stood in a corner of the room and contained a great diversity of books. On the topmost shelf, where they were unopened from one year to the next, were volumes by Charles Darwin, Henry Drummond, DeWitt Talmadge, Dwight L. Moody, and many others of similar depth and profundity. Most of these must have been acquired by inheritance for I am sure that my own parents would not have bought them.

On the second and third shelves one might find any of the books that we were currently reading and along with these were other such varied titles as *The Three Musketeers, Freckles, Hans Brinker, A Certain Rich Man, A Tour of the World in 80 Days, Black Rock,* and *The Friendly Road.* Here was a copy of *Lives of Our Presidents,* which included the just-elected Theodore Roosevelt. Here too was an old copy of *Uncle Tom's Cabin* which was modern fiction compared to our *New England Primer* hidden in the little desk drawer just below.

On the lowest shelf just above the desk compartment, for reasons of both convenience and appearance, were all of our really big books. Foremost of all, a giant in command of all the English, ready to give the final word to all these lesser tomes, stood *Webster's Unabridged Dictionary.* An enormous twelve-dollar leather-bound book, it was fully seven inches thick and so weighty with words that in my earliest days of its perusal Mother had to take it down for me and then later replace it on the shelf. But it was not this wealth of words that kept the book and me lying on the floor face to face for hours at a time. It was, instead, those thousands of illustrations that filled its final pages.

Here, arrayed before me page on page, were treasures from the entire world and from throughout all history. Here were colored pictures of the flags of every nation; the official seals of every state. Here seemed to be our planet's entire fauna—birds, mammals, fishes, insects from every corner of the earth. On display here were weapons from the arsenals of the ages and sailing ships from every sea. Webster and I spent many happy hours wandering through space and time and never left the living room.

I once read that some noted man of letters—I have forgotten who—was asked which three books he would choose to have with him if he were to be shipwrecked on some deserted island. His choices were, the Bible, Shakespeare, and a star atlas. In later years I was to come to a complete concurrence with these choices—with the one proviso that my Bible be the beautifully written King James Version, not the prosaic revision—but in those early years my own first choice would have been that book of childhood wonders—the dictionary.

Standing right beside the dictionary, and leather bound to match it, were the three large volumes of Ridpath's *History of the World*. These too afforded enthralling entertainment with their hundreds of illustrations depicting scenes from the building of the pyramids right down to the world of 1885, when the set was published. The works of the great poets were also well represented on this shelf with large well-bound editions of Shakespeare, Tennyson, Lowell, Whittier, and Longfellow.

We were never a family to just sit and talk; even less were we a family to just sit. We were a family of readers, even though our reading was often done in a vicarious way. Nearly every evening, especially when the weather was cold and we were all together around the stove, Mother would read aloud to the whole family. In this way we eagerly devoured all the works of such writers of that day as Harold Bell Wright, Ralph Connor, Rex Beach, Gene Stratton-Porter, William Allen White, Jack London, Zane Grey, and many others. These writers all were tremendously popular and their books were the best-sellers of that day—the books that all America was reading.

In recent years I have reread many of these same books which once had held our rapt attention. I found in all of them a rare and most refreshing feature. Every one of those books was a wholesome book. A book that could be read to anyone of any age without censoring a single word.

There was no public library in our vicinity at that time so we had to buy nearly all the books we read. While this may have curtailed the extent of our reading to some degree, it also gave us an even greater appreciation for each individual book. Each purchase we made was a distinct event in our lives and was made only after much careful deliberation. We learned to value books and to treat them with respect. For Christmas gifts we would buy books for one another. In this way we gave something which we all could enjoy and, as the average book cost only fifty cents, we had a long-lasting Christmas at a most reasonable cost. One could always be certain of getting to read precisely the book one wanted simply by giving it to someone else. If I just happened to give my sister, *Tom Swift and His Motorboat* I was also just as likely to receive one of the *Little Colonel* books from her.

We still have many of those books that we acquired in that era. My own first gift book was Robert Louis Stevenson's, *A Child's Garden of Verses,* dated, Christmas 1907. But the one I prize above all others is inscribed, "Christmas 1913, from Mother." This book, *Rolf in the Woods,* was written by artist-naturalist Ernest Thompson Seton. It is the story of how Rolf, a white orphan boy, was raised in the ways of the woods by the Indian, Quonab. To me the book was an open door into the world of nature.

Rolf became my hero and I sought, in every way possible, to follow in his moccasined footsteps. In order to pass him off as an Indian boy, Quonab had dyed Rolf's skin with a concoction made by boiling walnut hulls. I did the same. One Saturday morning, when all the family had gone to town, I got a small kettle from the kitchen and with it sallied forth to the front yard. There, beneath those walnut trees that Grandpa had stubbornly replanted, the trees that just three years before had framed a mighty comet, I found a little heap of walnut hulls that the squirrels had left there over winter. With these I soon had my little cauldron bubbling into a pungent inky stew which, when it had cooled sufficiently, I applied externally with generous abandon. Later, when I looked in a mirror, I hardly knew myself. Whether our Ohio walnuts were of a richer vintage than Rolf's Connecticut variety; whether my remote ancestor's rendezvous with a redskin had made me more receptive to the treatment, or whether I was just more thorough in the application I do not know but, at any rate, the over-all effect seemed considerably more African than Early American. Fortunately, it all turned out all right. After they recovered from the initial shock my parents thought the whole thing awfully funny and I was even excused from Sunday school attendance for two or three weeks.

Rolf slept each night in an Indian wigwam and cooked over an open campfire. I draped some discarded carpeting over some poles to make a tent and slept there on warm nights and I impaled bacon or ham on a long wire and broiled it over a small fire and then baked potatoes in the hot ashes.

The adventures of Rolf awoke in me an intense desire to learn all I could about the various forms of nature which, all my life, had been flying, crawling, blooming, swimming, and twittering quite unheeded all around me. Fortunately, among our books, we had a set of nature guides—pocket-size books with colored illustrations—and these were exactly what I needed, for our four volumes covered butterflies, wild flowers, trees, and birds.

I particularly favored butterflies and wild flowers. The butterflies and moths were easy to identify for I simply caught them in my home-made net and then compared them with the pictures in the guide. Wild flowers were even more obliging. I would find them in the woods, identify them from the pictures and then, whenever possible, I would dig up and bring home a specimen or two to plant in our rock garden for further observation. The birds, I soon found, were the least cooperative of all. I had no field glasses for close-up views of them and no bird could possibly have been stupid enough to mistake me for St. Francis and zoom in for a nearby landing where I could clearly see its markings.

I found, too, that the birds had still other complications in the matter of their identification. Now, in the order Lepidoptera, if you see one Cecropia moth, for example, you've seen all Cecropias. There is never the slightest doubt that Mr. and Mrs. Cecropia belong to each other. With the birds it often is quite a different story and any family resemblance between the male and the female and their immature offspring is purely incidental. Furthermore, to add to the complexities, birds, like bird watchers, like a seasonal change of attire. Many of the birds that we see heading south in the fall migration have "suffered a sea change" since last we saw them and are but a travesty of their springtime selves. Bird watching, definitely, is not to be taken lightly.

In my more reflective moments I came to feel a bit elated with the little things that I had found along the outdoor trail that Rolf had pointed out. It gave me the vague sense of belonging to some small inner circle to know that our box elder was really a misnamed maple tree; that a closer look among the Monarch butterflies sometimes showed up a masquerading Viceroy, or that the aerial acrobats of my summer evenings were not swallows but chimney swifts instead. I got the satisfying feeling that I too, like the old Soothsayer, now could read a little in "Nature's infinite book of secrecy."

But all the while that I was thumbing through those pages, reading a word here, a sentence there, the grandest chapters of that book lay there right before me, quite unseen, quite untouched.

6 Friendly Stars

WHO HAS NOT, AT SOME TIME OR ANOTHER, HAD A FRESH NEW IDEA suddenly strike him and then wondered in amazement, "Why did I never think of that before?" One May evening, when I was fifteen years old, I was standing out in the front yard when just such a new thought came to me. The night air was soft and warm, there was no moon, and all the brighter stars were shining. Something—perhaps it was a meteor—caused me to look up for a moment. Then, literally out of that clear sky, I suddenly asked myself: "Why do I not know a single one of those stars?"

Why indeed! I had known the Pleiades ever since my childhood days. I had watched two awesome comets wheel about the sun. I had seen the death-dive of a thousand shooting stars and on many a frosty night I had stood entranced as the ghostly fingers of the northern lights probed about the sky. Furthermore, I seemed to have an active interest in all the rest of the world of nature. But, until that night in May, I had given not one single remembered thought to the stars themselves. Why, I will never know. I can only wonder:

> When all the Temple is prepared within,
> Why nods the drowsy Worshipper outside?

Just as suddenly as the thought had come there came now the decision that I was going to learn those stars. In my mind I began to anticipate the fun I was going to have in doing this, for I still recalled the ecstatic moments which came to me with the identification of each new butterfly, flower, or bird. A couple of minutes later I was standing in the living room looking up at the varied titles in our high bookcase.

Seldom in the past had our home library failed me in any search for information. Our set of nature guides, Wood's *Natural History,* and Gray's *Botany* had served me well in all my problems with the flora and fauna of the farm. But the skies, it seemed, were a whole new world. We had not a single book about the stars.

Actually, this failure of the home fount of knowledge only cost a day's delay for I was now attending high school in Delphos and for the past several months I had been borrowing books from the new public library which was located less than a block from the school. Next day, as soon as school was dismissed, I parked my bicycle in front of the library, went in, and asked the librarian for a book about the stars. Had she had at her disposal all the titles in the Library of Congress she could have made no better choice than the book she brought me for it was Martha Evans Martin's classic work, *The Friendly Stars.* This book tells, simply and beautifully, the essential facts about the stars and constellations and, with diagrams, shows where to find them in the sky.

As soon as I reached home I hurried through the chores; then, after a quick supper, I started reading the new star book. There were two or three chapters of general information on the appearance of the night sky and on the rising and setting of the stars before it came to the chapters dealing with the identification of the individual stars. By the time I had finished reading these opening chapters it was growing dark. Quickly I leafed through the book until I came to the first star diagram. The caption beneath the diagram read, "Vega as she appears in the east." The drawing showed Vega as a bright star with a scattering of five fainter stars nearby—four of which formed an oblique parallelogram.

According to the descriptive text Vega, at that very hour in the month of May, would be rising in the northeastern sky. I took the open book outside, walked around to the east side of the house, glanced once more at the diagram by the light that came through the east window of the kitchen, looked up toward the northeast and there, just above the plum tree blooming by the well, was Vega. And there she had been all the springtimes of my life, circling around the pole with her five attendant stars, fairly begging for attention, and I had never seen her.

Now, I knew a star! It had been incredibly simple, and all the stars to follow were equally easy. Vega led the way to Deneb and these two made the base of a triangle with another bright star named Altair. Each star that I found would point the way to some new field and that, in turn, would light my path to yet another. Arcturus, the bright star that years later would turn on the lights for the 1933 World's Fair at Chicago, I found simply by following the curve of the Big Dipper's handle for a distance equal to the length of the handle. I followed that same curve an equal distance further and came to another bright star—Spica. Blazing red Antares was just as easy. In the middle of May, wrote Mrs. Martin, this star would rise in the southeast between nine and ten o'clock in the evening. On one of those mild mid-May nights I watched until a fiery gleam cleared the treetops of a distant woods and thus another star was mine.

I learned my stars as they were rising in the east and gave but little heed to those that then were dropping toward the west. I could see a few bright stars above the sunset sky but they were now retreating and would soon be lost from view. In less than half a year these would be rising in the east again and I would meet them then when they were coming toward me. In the meantime there was much for me to do. The skies were full of stars for me to learn.

The twenty brightest stars in the sky are grouped together in a class known as first-magnitude stars. Of these only fifteen could be seen from my latitude, the other five were southern stars. By the end of that first May I had made the acquaintance of five of the fifteen—all in the eastern sky. There would be no more additional bright stars until August, when two more would be rising in the east. So I decided that during this time I would learn some of the constellations. With the help of the diagrams in my star book I found it quite a simple matter to locate these legendary star groups in the sky.

I had learned the bright stars first and now I used them as unfailing guide posts to the constellations and the fainter stars. And in the learning of all these I was kept quite busy while the earth wheeled around the sun and brought new bright stars into view. With the exception of Vega, Arcturus, and Spica, which already were up in the eastern sky when I began, I learned all of these fifteen bright stars as they came up to meet me—over the eastern horizon. This method has a tremendous advantage, for when one knows that in the month of October a bright star called Capella will be rising in the northeast just at the close of day, one has but to look in that directiion to find it glowing in the early autumn twilight.

Learning the stars is pure delight and there are many pleasant ways to do it. No true star-gazer will fail to become familiar with the constellations and fortunate is he whose introduction to the skies comes to him through nature's eyes alone and not through any telescope. So few of those who use the eyepiece first ever get to really know the stars. There is a host of guide books to the stars and constellations now available and, no doubt, all of them have merit. Some of these, I feel, stress too much the old mythological figures of the constellations and thus lead one to look for things that are not there. Still others make an acquaintance with the stars a most tremendous task when really nothing could be simpler. At the opposite extreme, there are a few that tend to oversimplify and offer many shortcuts to a knowledge of the stars. One also can make a hurried dash through Yosemite Valley—and miss so much in doing so.

To me, the least satisfactory way of all to learn the stars would be through the eyes of another. The organized "star-party," or the constellation study groups in which someone points out the various stars and constellations are pleasant social affairs but they make it all so effortless that the lesson seldom sticks. It is like taking a guided tour to see some wonder of nature when one could, just as well, have the incomparably greater thrill of being its discoverer.

It took me about a year to become acquainted with the stars. This may have been a longer apprenticeship than some would care to serve but I have found it well worthwhile for in the end I had much more than just a mere assortment of names and places in the sky. Each star had cost an effort. For each there had been planning, watching, and anticipation. Each one recalled to me a place, a time, a season. Each one was now a personality. The stars, in short, had now become *my* stars.

One of those nights of that first year with the stars was, for me, a memorable occasion. I still had the little tent of my Indian days out under the maple trees and on hot nights that summer I often slept there on a canvas cot. One sultry night in early August I decided to have a preview of the stars which the yearly circling of the earth about the sun would bring into my evening skies during the coming fall and winter seasons. I set the family alarm clock to call me at 4 A.M. and crawled into my tepee a little earlier than usual. At four o'clock when the alarm sounded it was still quite dark and a bit chilly. I didn't bother to dress but wrapped up in a blanket and stepped outside.

The moon had set and the sky was beautifully clear. In the

north were the circumpolars—the stars that never drop below the horizon. With these I was familiar. From the polestar the Little Dipper dangled straight down as from a nail in the north wall of the sky. High up above me was Cassiopeia while, far down below, the Big Dipper skirted the horizon so closely that the star in its handle-tip sometimes seemed to catch in the distant tree-tops. Now low in the west were Vega, Deneb, and Altair—stars of the east in my evening sky. Halfway up in my predawn sky were my old friends, the Pleiades. All the other stars were complete strangers.

My early morning meeting with all these unknown stars had been deliberately planned. I wanted to see these stars as a total stranger sees them. I wanted to see them as an earlier age of mankind had seen them—an age that found in these same skies strange figments of their folklore. Just how, I wondered, would this unknown starland strike MY fancy. What weird figures, beasts, and monsters would I see through my still-unbiased eyes.

In looking about I saw six or seven stars, unknown to me, which seemed to be in the first-magnitude class. There was a lone, white star, about as bright as Deneb, in the southwest. A short distance east of the Pleiades was a reddish star which formed part of a letter "V." On the eastern horizon were two stars of nearly equal brightness with another brighter one high above them, while in the southeast hung two bright stars with a diagonal row of fainter stars between them.

I gazed at all that bright array of foreign winter stars and tried to see in them some elements of crude design. I tried to see those stars as groupings that might make some legendary figure or perhaps some ancient god. Then in my mind's own time machine I traveled back five thousand years and tried looking through the eyes of some Chaldean shepherd, some Euphratean nomad, some sailor from Phoenicia. But it was all of no avail. Only twice did I see anything at all. In that lone southwestern star I could see a Cyclopean monster and of a looping trail of faint stars in the south I made a writhing snake. But in all the rest I could find no fanciful resemblance—it was just a careless scattering of stars to me.

On other, later nights I found that I just lacked imagination.

My one-eyed Cyclops was the Southern Fish and my slithering serpent was Eridanus, the River. And in those other random stars, where I had failed completely, earlier eyes had seen a mighty Hunter and a charging Bull, a Charioteer and Gemini, the Twins.

As a constellation maker I had been a dismal failure. But I consoled myself with the thought that maybe I had been working under a handicap. I wondered if any of those ancient artists of the East had tried to draw their fanciful sky figures while draped in a blanket and with their bare feet drenched in the chilly dew of early dawn.

Actually that preview of the stars was a big success! I had wanted a night to remember and I got it. I wanted a memory picture of an unknown sky and I still have it. I saw the winter stars that night as Balboa saw the chartless Pacific, as James Cook saw the South Sea Islands, as Champlain saw the frontier wilderness. And just as all the sights they saw are commonplace today, so was my first meeting with the winter stars the forerunner of a thousand later alarm-clock meetings with the early morning stars of every season.

All through the months of that first summer *The Friendly Stars* was the companion of my starlight nights. Every two weeks I would return it to the library—and bring it home again. Naturally, it got a lot of outdoor usage and its pages bore the prints of fingers that were far from clean and, by summer's end, it did not have that fresh, new look it had when I first took it over. I have always felt that a good book should bear some signs of usage. My friendly star book qualified in every way.

Just a few years ago a thoughtful librarian, knowing of my long-continued interest and how it had been nourished in my earlier years, presented me with a copy of *The Friendly Stars.* It had a beautiful dark cloth binding with the title lettering all in gold. In a momentary daze at this gesture of good will I opened up the volume just at random. It opened on the diagram of Vega in the east. Amazed, I tried it again and Vega reappeared. My suspicions aroused, I turned over a few pages and saw what filled me with delight. Between those brand new covers it was my same old *Friendly Stars*—complete with all the finger smudges I had left there in my youth! Then, recalling one final bit of evidence, I turned quickly to the chapter on Altair. There it was—the dark stain made one summer night when the book was closed on a May-fly.

With the book back once more under my arm I went down the wide stone steps of the library and then along the walk leading to the street. Out near the end the same silver maple crowded even closer up against the slightly tilted flagstone. But this time there was no white bicycle leaning up against the tree, waiting to take me out to the farm and my evening chores.

I put the book down on the seat beside me and drove on home. So many changes had taken place, so many things had come and gone since that schoolday when the book and I first went down the library walk and rode away together. Great battles had been fought. Once mighty states and empires had been swept away. New boundaries had been established; new frontiers had come and gone. But I knew that in my star book I would find no changes in its skies. The timid Hare still crouched unmolested beneath the mighty Hunter, the Pleiades still dispensed sweet influence just as in the days of Job, and each spring Vega and her attendant stars could still be found above the plum tree on the farm.

I had seen new stars from time to time during these years but soon these faded out and only for a little while did they disturb the constellations. I had watched a dozen comets, hitherto unknown, slowly creep across the sky as each one signed its sweeping flourish in the guest book of the sun. These too had come and gone. Along with all these transient visitors many of my own allotted hours had also come and gone since that May night when I first spotted Vega in the east. I did a rapid reckoning. That star book there on the seat beside me had a lot to answer for. All the hours of two entire years had been spent in gazing at the skies that it had pointed out to me.

Each year when May returns and brings her balmy nights, and Vega with her trailing lyre strings rises in the east, I still can hear, no matter where I am, the telephone wires along the road all humming background music for the rhythmic treble chanting of a distant choir of tree frogs. And when Orion stalks across midwinter skies his giant steps are followed by the sounds of squeaking snow and jingling bells from country sleighs that vanished nearly fifty years ago.

Come to think of it, maybe the stars came out for me at just the right time, after all. Impressions sink in deepest when the clay is warm and fresh. The year, that May, was in its spring and I, at fifteen, was in mine.

7 The Strawberry Spyglass

WE MUST ALWAYS HAVE GROWN STRAWBERRIES. IN MY OWN MIND, at least, it goes back quite beyond remembrance. With us it was not just a row or two somewhere in the vegetable garden, it was a small commercial enterprise involving, perhaps, a couple of acres of ground.

It seemed to me then that we picked berries all summer long, for in one's early years time passes slowly. In reality the berry season never included more than the thirty days of June and even this extent was attained only through a careful selection of early, midseason, and late varieties.

Strawberries, as we grew them, were a constant care for about eight months out of every year. In April the layer of straw that had covered the plants over winter was raked off and placed between the rows where it served to keep down weeds, helped keep the berries clean by preventing dirt from splashing on them during rains, and it also provided a most welcome cushion for the knees of the berry pickers. On many an occasion it also became a quickly-available frost cover. Frost was always a threat during the month of May when the plants were in full bloom and we often had to rake the straw from between the rows back over the plants again on short notice when an advancing cold front brought clearing skies and falling temperatures on late afternoons.

In view of the strawberry's year-round association with straw, both as a mulch and as a winter covering, I was surprised to learn, in a recent reading of Samuel Fraser's book, *The Strawberry,* that the plant derived its name from its peculiar habit of growth. After the fruiting season the plant increases by putting out a number of runners which, on contact with the ground will root and form new plants. These runners and new plants have the appearance of being strewed or, in the ancient form of the word, "strawed" about the mother plant, and thus the name strawberry resulted.

We set out a new berry patch every year, usually in May and these new plants required constant care throughout the growing season. Every year Dad tried out some of the new varieties that were constantly appearing. When I recall some of those old-time strawberries we used to raise I respectfully take off my weather-beaten straw hat to the United States Department of Agriculture and to the commercial plant breeders for the mighty changes they have wrought. As I remember it now, our old reliable early variety, Warfield, was but little larger than a good-sized cherry and it seemed to take forever and a day to pick a quart of them. Today's berries are more than twice the size of the old varieties and a good picker can fill a quart basket of them in just a couple of minutes.

June was a busy month for the whole family. In one way or another we were all involved in the berry business with the exception of my older brother, Kenneth, who ran the other farm operations during this critical period. Corn requires continual cultivation during this month to keep weeds from getting a start, and all day long, every day, he rode the two-horse cultivator slowly up and down the long rows of corn. My mother was in supreme command of the small shed we called the Berry House where she served as both timekeeper and inspector. Here she punched the pickers' cards each time they brought in a carrier of eight quarts of berries and, if the picker was young or inexperienced, she also sorted through their quarts for any possible defective berries. My sister, Dorothy, and I were on the picking force while Dad, of course, supervised the entire operation.

Strawberries were Dad's pride and joy and he spared no pains in raising the very best to be found anywhere. I recall two trips that he made by the interurban train to the R. M. Kellogg Company of Three Rivers, Michigan, in order to look over their propagating fields and select new varieties to try out here at home. We always marketed our berries in brand-new quart baskets. But merely being new was sometimes not enough and Dad shopped around until he finally found a factory in Berlin Heights, Ohio, which put out baskets made of a very clean white wood that Dad said was buckeye. We all felt that these must be tops—Ohio being the Buckeye State.

We usually started picking in the morning at about nine o'clock after the heavy dew was gone, and we finished for the day at about two or three o'clock in the afternoon. Then Dad would hitch up the horse to the Berry Wagon and take the load to town and deliver them to the various grocery stores. This Berry Wagon, as we called it, was another example of Dad's rural artistry. He had taken an old spring wagon and built a special body for it which he had designed to hold twelve twenty-four quart crates of berries. On top of these bottom twelve crates another similar layer could be added if necessary. He painted the body a light cream color and decorated it in black with a large and intricate trademark on either side, while on the tailgate, in flowing script, he neatly lettered:

Stanley W. Peltier's
Strawberries

He firmly believed that a good product should be properly identified and each crate was given a printed label and each quart basket also bore his name stamped in dark blue on its white wood rim.

During the summer months we often drove to town in the Berry Wagon on Saturday evenings. This was a sort of shopping spree—an opportunity to buy a pair of overalls, a window shade, new shoes, a thirty-five cent steak for Sunday dinner—the things we couldn't buy from Tobe Luttrell's huckster wagon that stopped in front of our house and rang its cowbell every Thursday morning.

I always liked these trips to town in the Berry Wagon. I got to ride in the back end of the wagon, with the tailgate open and my feet dangling down below. We usually drove to town over what we called the Lower Road, which in later years became the Lincoln Highway. Only occasionally did one see an automobile traveling this dusty road but in Delphos the authorities were taking no chances. On the north side of the highway just where it entered the town there stood a sign which read: SPEED LIMIT—8 MILES PER HOUR. As we drove by this I sometimes cautioned Dad that maybe he ought to slow down just a little.

Occasionally, on these Saturday nights I would attend a "picture show." The movies had come early to Delphos, the first performance that I saw was *Ali Baba and the 40 Thieves* in about 1907. I do not recall any of the actors. At one time there were three movie houses in operation here—one being an outdoor fair-weather establishment. On our Saturday nights on the town for the sum of five cents—if I had it—I could listen to a live three-piece orchestra while I watched the histrionics of such early notables as King Baggott, Mary Pickford, Owen Moore, Fatty Arbuckle, and Francis X. Bushman. Or, across the street, to the accompaniment of a player piano I could sit captivated as Tom Mix or William S. Hart corralled the outlaws and then rode away into the sunset.

Uptown parking on Saturday night was just as much a problem at that time as it is today except that it was not a question of finding an empty space along the curb, but instead, an empty place at a hitching rack. By coming to town early we could usually find a vacant place on East Fourth Street just off Main Street. This was such a convenient location that even Mother ignored the proximity of the busy saloon right across the street on the corner. When we could hitch here we were only a very short distance from the Masonic Hall and also from Charlie Ray's grocery store. Charlie was a steady strawberry customer of ours and we always bought most of our groceries from him. His store made, in addition, a most convenient family meeting place where we gathered, one by one, at about ten o'clock and then started out for home. As we left, carrying our big paper bags of groceries, we walked a few doors to the north and tapped five times on the heavily curtained window of the Masonic club room where Dad had spent his evening playing billiards with some of his faithful brothers of the green felt lodge. Then we all headed across the street back to the wagon.

Going home I never insisted on riding with my legs hanging over the tailgate. I was glad now to lie down flat on the floor of the wagon on a blanket brought along for just that very purpose. As we were homeward bound on one of those clear summer nights we were passing along a stretch of road with fields of tall ripening wheat on either side when I heard Mother remark to Dad that it was hard to tell just where the lightning bugs stopped and the stars began. I raised my head up from the blanket and took a drowsy look. The low sides of the wagon cut off all view of the ground below. I could see only the bright band of the Milky Way crossing the sky high above me and the twinkling and blinking stars all around me on every side. I lay back on my bumpy bed and watched for a long time as we sped on through the star fields of space. When I awoke we were back on earth beside the barn.

Along in my mid-teens I was to discover that strawberries and stars are most compatible and that such odd things as telescopes are sometimes measured by the quart. But it was long before this that I first became aware of the pecuniary possibilities of the berry patch.

When I was eight years old I picked berries to earn my first dollar and it took me nearly the entire month of June to do it. At that time we were only paid one-cent-per-quart for our efforts and in my case those efforts certainly shattered no records. There were just too many things happening on a bright June day for me to waste my life away in common toil. There were fleecy cumulus clouds that had to be watched to see if their projected shadows would cross the berry patch or miss it. There were red squirrels running up and down the old rail fence and chattering in the walnut trees and these all needed my attention and high up in the blue sky overhead there was always a turkey buzzard or two that deserved a lot of wondering about how they could manage to sail around for hours and never have to flap their wings.

It seemed, too, that there were always some important comments to exchange with the two or three other pickers of my own age, and I can recall still other interruptions that were occasioned by an overassiduous quality test of our product. Sometimes there were other distractions right at hand. Often, as I was picking, I would hear a sudden fluttering sound and a little ground sparrow would fly up from its nest in the row just a few feet ahead of me. When I eventually picked my way to the spot I would find, deep down among the berry plants, a neat little structure made of woven grasses interlaced with a material quite unknown to the farm birds of today—horsehair. I was always careful not to touch the eggs in the nest for I had been told that the mother bird could tell that they had been touched and would abandon the nest. Robins were a real pest in the strawberry field. They always selected the ripest and largest berries, took a few choice bites, then went on to find another. Fortunately, the starlings, which now have made smalltime operators out of the robins, did not move in until the mid-thirties.

I spent that first dollar in one fell swoop! It went for a year's subscription to *The American Boy,* a truly wonderful magazine which, unfortunately, is no longer published. I still recall the excitement of removing the wrapper from that first issue of July 1908. Its colorful cover depicted a boy on horseback at the head of a parade while above him he held an immense flag with forty-six stars.

Between the covers of that magazine was a whole new world. I had never known before that there were so many things to be interested in. There was a page for stamp collectors, another on coin collecting, still another on taxidermy. There were departments devoted to nature study, photography, and even model airplanes, although the flight at Kitty Hawk was but five years gone and the English Channel would not be flown for yet another year. There were thrilling stories of adventure, of travel, school life, and historical novels. Later issues were to present, in serial form, the amazing adventures of a very fat boy named Marcus Aurelius Fortunatus Tidd, or Mark Tidd, for short. The story was written by a promising young author by the name of Clarence Budington Kelland.

I continued to subscribe to *The American Boy* for many years and kept all the back issues (I save things), but one winter a neighbor borrowed them and they never came back.

One day, near the close of my second year in high school, I chanced to see a small telescope in one of the wall cases in the physics laboratory. The instructor allowed me to examine the instrument a little more closely. It was a collapsible drawtube affair of the type one often sees pictured in the hands of the old sea captains. I had a great yearning to see through it and to this end I obtained an audience with the principal and asked permission to borrow it to look at the moon and stars. My request was politely but firmly refused. I had rather expected this and, though I was somewhat disappointed at the time, in the end it was all for the best. This refusal actually was just an early example of the well-disguised blessings that so often have come my way for now I was determined to have a telescope of my own.

All this happened, of course, long before the age of telescope making, so I did not have to decide whether to buy a telescope or make one, nor did I have to weigh the relative merits of a refracting telescope versus one of the reflecting type for small reflectors were still completely unknown. I finally found two companies that offered small telescopes for sale. A mailorder house listed a 3-inch Bardou refractor for sixty-five dollars and the A. S. Aloe Company, of St. Louis, advertised a 2-inch telescope for eighteen dollars. Again, I wasted little time in arriving at a decision for sixty-five dollars was completely out of sight. I did not have even a start on the eighteen dollars but my hopes were high for June, the month of strawberries, was just around the corner.

At the berry patch the picking rate had gradually increased over the years until it was now two-cents-per-quart. Even more important than this, I was now a picker with a purpose. The result was a fine example of the effect of incentive on take-home pay, for before the month was over I had picked my nine hundred quarts and my eighteen dollars were speeding toward St. Louis.

According to my figures I should have my telescope in four days—five days at the very most. My order should reach St. Louis in two days. There, a clerk at the Aloe Company would read my order, place a telescope in a box, address it and mail it back that day and in two more days it should arrive.

It didn't. Nor did it come the next day, or the next, or the next. I began to wonder if the company was on strike, or if their clerks were all on vacation, or if, maybe, the shipment had gone astray. Then I thought of still another possibility. Art Moon, our mail carrier, made his rounds on a motorcycle. He carried all the first class mail in a wire basket attached to the handlebars in front of him, but the parcel-post packages, such as mine would be, were stowed in cavernous open saddlebags that hung on either side of the rear wheel. Maybe the motorcycle had hit a bad bump in the road and my package had jolted out. Maybe someone else was using my telescope.

On the morning of the ninth day after mailing my order I was hoeing corn in a field not far from the house. Art always made our stop at just about eleven o'clock and for the past week I had contrived to be at the end of the field which had a clear view of the mailbox at that time. Promptly, on the hour I heard the familiar chug of his single-cylinder Excelsior as it passed along the road at the end of the cornfield. Leaning on the handle of the hoe I watched expectantly as Art coasted up to the mailbox and stopped. He took out a couple of letters, dropped them in the box and closed the lid with a bang which clearly sounded like, "No telescope today." Then for what seemed like a long time he worried with the strap that secured the remaining letters in his wire basket. Finally, almost as an afterthought, he half-turned on the saddle, pulled a longish package from the right-hand saddlebag and laid it on top of the mailbox.

With a roar Art was off in a cloud of dust. With a yell I too was off, leaving behind me a slowly toppling hoe and a swirling wake through the knee-deep sea of corn.

Plopping on the ground right beside the mailbox I hastily removed the outer wrapping of corrugated paper to find inside a round case of heavy cardboard. I pulled off the cover of this case and there, wrapped in tissue paper, was my telescope—a beautiful thing of black pebbled leather and shining brass. Gingerly I pulled open the drawtubes until all four were fully extended. I then removed the brass cap that covered the lens, pointed the telescope toward Grandpa's house across the field and, with the feeling that this was one of life's great moments, I looked in the eyepiece. I saw only a blur!

In that first frenzied hour I learned many things about telescopes. First of all I learned that they need to be focused. That an adjustment of the distance between the large lens and the eyepiece—made by sliding in or out the smallest drawtube carrying the eyepiece—is necessary in order to make the image sharp and clear when seen in the eyepiece. I learned that this adjustment varies with the distance of the object that is viewed, that the more remote the object the nearer must be the eyepiece to the lens. And to my great delight, when this adjustment had been made, I learned that Grandpa had flies that sunned themselves on the siding of his house and dark green moss that grew between the red bricks of his chimney.

At noon I showed my spyglass to the family and they all thought that it was quite wonderful. After looking at a couple of distant houses Mother remarked: "It's a good thing everybody doesn't have one of these. It'd be worse than our party line." Dad played with it for a long, long time before he finally took his hoe and left for the berry patch and I got it back again.

About the middle of the afternoon I drifted back to the cornfield but I took the scope along. I would hoe a round, then I would sit one out, drawing a bead on everything in sight. I watched the birds in the tops of the walnut trees along the road. I spied on the cows in a distant pasture, on the pigeons lined up on the peak of our barn roof, and, right above them, on the gilded horse that forever chased the wind around our southmost lightning rod. Finally, remembering, I looked at our church a mile away and could count the shingles on the roof. According to the story in my *Short History of Natural Science* a church had been the first object seen through any telescope. Three centuries before, the children of a Dutch spectacle maker, while playing with two lenses, happened to line them up, in focus, with a distant church.

My new telescope had a lens 2-inches in diameter with a focal length of 36-inches. It came equipped with two eyepieces, of 35X and 60X magnification. The instrument was of French manufacture, well constructed of heavy brass and with smoothly working drawtubes. Optically, it was a terrestrial telescope for it had an erecting lens permanently mounted within the tube so that objects viewed on this terrestrial ball would not suffer the indignity of appearing upside down. An astronomical telescope dispenses with this erecting lens, which naturally absorbs some of the light it receives and, as a result, everything seen in such an instrument appears reversed, top to bottom, right to left. To the astronomer this reversal is of no consequence but in using such instruments in later years I was sometimes to find it quite disturbing to visitors to see a crescent moon facing one way with the naked eye and exactly the opposite way in the telescope. No doubt some of them figure that the whole thing is a fake.

The sky was cloudless on the day my telescope arrived but nevertheless I watched anxiously throughout the long hours of the afternoon. Even before it was dark enough for the stars to show I was ready and waiting for them. Weeks before I had decided that my first look would be at Vega. It had been she who started out with me and it was fitting that she now initiate my spyglass. And so, before the darkness had completely settled for the night, her blue-white dazzle scintillated in my field.

This was followed by an hour with the moon, with Mars, and with a number of the brighter stars. Everything I saw that night amazed and delighted me and each new sight assured me that my starlight nights would be busy nights for a long, long time to come. My very first look at Vega, however, revealed that I had completely overlooked one most essential adjunct to the well-tempered telescope. It had no mounting or support of any kind to hold it steady.

Just how such sundry sea dogs as Captain Cook, Long John Silver, and John Paul Jones ever managed to stand on a pitching upper deck and focus one of these freehand spyglasses on a distant sail is quite beyond me. Even with both my feet firmly anchored on solid ground the stars all made mad gyrations across my field of view and I attained a measure of success that first night only by resting the telescope against obliging tree trunks, fence posts, and the corner of the house. The next day I built a mounting.

It is well that I once took a photograph of this mounting. Without this graphic proof I would be hard put indeed to furnish acceptable evidence that anything like it ever really existed. It proved to have a number of unique features and so, for the benefit of anyone who might wish to construct a telescope mounting that does not slavishly copy the conventional models now on the market, here are a few helpful hints on how I made it.

First, I selected a 6-inch diameter fence post that had been retired from active service and cut it to a 3-foot length. Then, from a two by four I cut three pieces each 2½-feet long, making the cuts on a 45-degree angle. These were spiked to the lower end of the post, spacing them 120-degrees apart. Any spikes that happened to bend were left that way for they helped give the mounting a certain handcrafted look and they also provided added ballast. When these three legs were placed on the ground they held the post in an essentially upright position.

On top of the post I now placed an old discarded grindstone that we had out in the barn (we save things). This stone measured twenty inches in diameter, was two and one-half inches thick, and weighed about thirty pounds. Then, from what was left of the two by four, I cut two pieces each one foot long and spiked them together to form an "L." A quarter-inch hole was then bored in the center of the base of the "L" and this hole was then centered on the grindstone and a large spike inserted through it and driven into the wood-plugged hole in the grindstone and on into the top of the post. This permitted the "L" to turn freely in a complete circle with the grindstone as its base. A notch was cut in the upright arm of the L and the offset telescope support which I then made was fitted so that it pivoted up and down on a bolt through this notch.

What could have been more simple? Yet some people will pay out good money for a ready-made mounting that has no individuality whatsoever. Still others will enlist the aid of a skilled mechanic to help them make one. Here, in less than a day's time and for absolutely no outlay of cash, I had a truly different mounting that held my telescope as steady as a rock. It weighed less than seventy-five pounds and I could move it from place to place around the yard to avoid the trees and house with no more physical exertion than one can readily imagine.

We who are not perfectionists are vastly tolerant to faults, particularly our own. But, nevertheless, my mounting did have its good points. It stood out unprotected the whole year round with never a complaining creak and it weathered gales without a tremor. It served me well, just as it was, on virtually every clear night for nearly three-and-a-half years. It was the only mounting the 2-inch ever had.

8 Sky Exploring

DESPITE ITS DIMINUTIVE SIZE A 2-INCH TELESCOPE IS FULLY CAPABLE of doing serious work. A quality scope of this aperture will reveal a representative example of every major class of celestial object that can be seen with the very largest instruments.

During the tenure of the 2-inch I tried to see for myself everything that was listed as being within the range of a small telescope. Without too much difficulty I was able to locate all of the then-known planets and I admit to a most tremendous thrill when

I added the outer planets Uranus and Neptune to my slowly expanding universe. To capture these two planets I simply drew up for each a sky map or chart that showed all the stars visible in my telescope in the region of the sky through which the planet was moving at that time. With the chart before me it was then a matter of comparing it with the actual sky from night to night until one "star" of those that I had drawn disclosed a slow shift in its position. By this I knew that I had found a wandering brother of the earth.

Once I knew its position I discovered that Uranus was quite easily visible to the naked eye. In the telescope I could detect its greenish hue, its absence of flicker, and, I thought, a very tiny disk. Neptune, on the other hand, looked just like any eighth-magnitude star but it gave me quite a "professional" feeling to know that I had found the planet, just as the astronomer Galle had first sighted it seventy years before, by checking it against a sky map. It was very satisfying to feel that the 2-inch and I had now traveled to the outer limits of the solar system. This feeling vanished fourteen years later when the planet Pluto was discovered. In this same slow fashion, just as they had first been found a century before, I later brought to earth the four brighter minor planets, Ceres, Pallas, Juno, and Vesta.

Jupiter was a joy to observe. There was always something doing; changes were taking place from night to night, even from hour to hour. His four bright moons, his cloud belts, and his polar flattening were all easily observed in my 2-inch though I find no mention of his Great Red Spot in any of my records of things seen so it probably was inconspicuous during those years. Saturn was at his very best that summer, being well above the southern haze, and tilted enough so that I could see the underneath side of the ball and rings with a small crescent of the northern hemisphere showing above the broad sweep of the outer ring. The sharp black void known as Cassini's division, which lies between the rings, could be seen on good nights as could Titan and Japetus, its two brightest moons. Occasionally, during daylight hours, I watched sun spots through my filtered eyepiece and followed their changing aspect as they moved from day to day across the solar disk.

I never tired of looking at the moon and in those early nights of watching with my 2-inch glass I often thought of Galileo and his tiny telescope. When news of the invention of the telescope in Holland finally reached Galileo in Italy he immediately set about making one of his own. For this purpose he purchased two spectacle lenses from an optician. One of these, of convex shape like a reading glass, became his objective lens. His eyepiece was just the opposite in curvature, or concave. The two lenses were then fitted into the opposite ends of a leaden tube of the proper length to focus on distant objects. Then, just a little more than three hundred years before, in the spring of 1609, Galileo turned his tiny homemade "optic glass" upon the moon. It was man's first skyward look through any telescope. As he related it:

> I betook myself to observations of the heavenly bodies; and first of all, I viewed the moon as near as if it were scarcely two semi-diameters of the earth distant. After the moon I viewed other heavenly bodies, both fixed stars and planets, with incredible delight.

I feel quite sure that I first viewed the moon in my small scope with just as much incredible delight as Galileo did in his. It is true that I had seen photographs of the moon and therefore had some vague idea of what its appearance would be like, but I still was wholly unprepared for all the wonders which I found on that first night as I explored the lunar surface. No photograph has yet been made which is not cold and flat and dead when compared with the scenes that meet one's eyes when the moon is viewed through even a small telescope.

From an aesthetic sense, at least, it takes a small telescope and low powers to do full justice to the moon. While it is often possible to utilize the very highest powers of the scope to ferret out some small detail of cleft or craterlet, it is only when the entire moon fits comfortably within the field of view that she is at her dramatic best. Only when one sees some empty space about her does she seem to float suspended in the sky like the neighbor world she is. This is the way I always saw her in my 2-inch telescope.

I spent many of those early nights in wandering aimlessly about the moon. I followed the advancing sunlight all the way across her face. I descended into craters by the score—Plato, Eratosthenes, Copernicus, Tycho, across majestic Clavius, and down the blinding wall of Aristarchus. One night I walked across the strange and violent gash of the Alpine Valley and then I climbed a tortuous trail from peak to peak along the sweeping range of Apennines. I rested briefly in the long black shadows of Pico and Piton, whose towering monuments rise starkly from the level surface of the Sea of Showers.

From night to night in its march across the sky I watched the moon grow from a slender sliver in the west to a full-orbed globe above the eastern treetops. On some of those nights I saw strange lights offshore in the sea of darkness that ebbed before the line of sunrise creeping out across the moon. An hour later these weird points of light had turned to mountain peaks and crater rims as the rising sun slid down their slopes toward the still dark plains below. On the seventh night of the lunar month I explored the Grand Canyon of the moon—the Hyginus Cleft, and one night later I rode the Lunar Railway from terminal to terminal across the floor of Mare Nubium.

Throughout these nights of discovery and exploration of the moon one question kept recurring to my mind. Why had I been denied all this until my school years were so nearly spent? Why had it not been made a part of the growing up of every youth? I had been taught the rivers, the seas, the mountains of every continent on earth. I knew the capitals of every state and country in the world. And all this time, right above me, the "geography" of a whole new world had been turning, page by nightly page, and no one had opened up the book for me. This was not a negligence peculiar to those times—it still exists. In later years with other telescopes I was to show the moon to thousands of visitors of all ages and not one knew the name of a single mountain range or crater on the moon!

When Christmas came that first star-gazing year my mother

gave me a small book entitled, *A Field Book of the Stars,* by William Tyler Olcott. Just as *The Friendly Stars* had been it was precisely the book I needed for it began its story right where the first star book had stopped. This was a lucid guidebook. There was a separate page for each of the constellations, large and small, which can be seen in this latitude and the opposite page, in each case, describes the constellation in detail and calls attention to its points of interest. I especially liked the way the constellation diagrams were drawn. They seemed to have been made while actually looking at the sky and the principal stars were connected by lines in a way that made them easy to recognize. For example, Bootes was depicted as a kite-shaped figure in the book and, as anyone can see, it is a kite-shaped figure in the sky. There was no attempt to show Bootes as the Herdsman or Bear Driver that mythology says he is.

The individual star names, as well as the Greek letters, were given for the various stars in each constellation. But the feature that I appreciated most of all was that while the constellation figures were drawn up for the naked eye, they also showed the locations of many of the objects of special interest in a small telescope, such as star clusters, nebulae, double stars, and variable stars. I had known that there were such things as these but until the field book pointed the way, I had no idea where to find them in the sky.

With the open book as our chart and compass the 2-inch and I now embarked on a treasure hunt among the stars. Night after night, page after page, constellation after constellation, we saw each cluster, nebula, and double star whose brightness came within our range. The Great Nebula in Andromeda, The M13 cluster in Hercules, the Dumbbell Nebula and, far to the south, the Trifid Nebula in Sagittarius; these and a host of others I thus met for the first of many meetings. In Orion's Sword the poet's "single misty star" was transformed by my glass into four tiny sparkling suns enmeshed within a swirling cloud of gas. The Pleiades became, one night, not seven stars but fifty, and Alcyone, the brightest one, revealed his three nearby companions.

And there were many other stars—Rigel, Mizar, Porrima, Castor, Polaris, Albireo, and Cor Caroli—to name but a few of the brightest, who felt the power in my three-foot magic wand and thus divulged, sometimes reluctantly, that close beside them, all but lost within their glare, another star was hiding.

My exploration of the skies took many months for there were many things to see and each new season brought new vistas and new wonders into view. It was well that I thus learned to chart my course among the stars for before the Christmas field book had been mine a week I was fitting for a voyage which would take not months, but years. A voyage which, indeed, is under full sail yet today.

My sailing orders for this celestial cruise came to me in the form of a footnote which author Olcott had added at the very end of *The Field Book of the Stars.* It read:

> Many readers of this book may be the fortunate possessors of small telescopes. It may be that they have observed the heavens from time to time in a desultory way and have no notion that valuable and practical scientific research work can be accomplished with a small glass. If those who are willing to aid in the great work of astrophysical research will communicate with the author he will be pleased to outline for them a most practical and fascinating line of observational work which will enable them to share in the advance of our knowledge respecting the stars. It is work that involves no mathematics and its details are easily mastered.

What teen-age star-gazer with a new and eager telescope could resist an eloquent appeal like that? I had not the slightest idea of what the author meant by "astrophysical research" but it sounded so lofty and important that whatever it was I wanted it. I immediately wrote Mr. Olcott asking for further information. Within a few days I had his reply—a long handwritten letter explaining that the research referred to was the systematic observing of variable stars.

William Tyler Olcott, I eventually learned, also was an amateur star-gazer, but he was one who had practically made a career out of it. He had received a degree in law from Trinity College in Connecticut but had never actively practiced his profession. In 1905 he became interested in the stars and his *Field Book of the Stars,* published in 1907, was the first of several books which he wrote on stars and star lore.

Through frequent contacts with Director Pickering of Harvard College Observatory, Olcott became aware of the profession's great need for more data on variable stars of long period. Believing that this was something that the amateur telescopist could well do, in 1911 he brought together seven interested observers who thus became the nucleus from which soon developed the formal organization known as the American Association of Variable Star Observers, or AAVSO for short, which was dedicated to the systematic watching and the reporting of the doings of these mysterious stars. Mr. Olcott had been elected the life secretary of the organization that he had founded and, included in my letter from him, was an application for membership in the AAVSO.

Alas, one of the requirements for active membership listed on the application blank was that one possess a telescope of 3-inch aperture or larger. As mine was but a 2-inch this was indeed a blow and I saw my rosy dreams of being an astrophysical researcher fading away by inches. For a couple of days I did some solemn soul-searching and then I finally returned my application with my telescope properly listed as a 2-inch. Too late, however, I noticed that my figure "2" started out with an extra curlicue at the top and actually looked more like a "3." Since it had been done in ink I would only have messed it up in changing it so I let it go as it was and thus, before the new year was well under way, I found myself a bona fide member of the AAVSO.

That winter of 1917-1918 was the coldest on record in this locality. Early in January the temperature fell to twenty degrees below zero. But, cold as it was, there were still the usual old-timers, including Grandpa, who assured us that our weather was mild compared to the winter of eighteen-something, when the rivers froze solid and when they threw boiling water up in the air and it came down solid ice.

Schools were closed and there was no rural mail delivery for a week or more as all the country roads were filled with drifted snow. At home our entire world had shrunk until its radius was now no more than fifty feet, and its two continents—house and barn—were connected by a single narrow trade route through a sea of dazzling crystal. But our little world was a self-sufficient one. We could have lived here comfortably and, I am sure, quite happily through weeks of wintry isolation. In town, during this same period, frozen water and gas lines, together with a war-induced coal famine, brought hardships to many. Here on the farm life went on just about as usual. Our woodshed was full of fuel for the stoves and our two wells, one near the house, the other in the barn, never failed us. We primed these pumps once each day with a teakettle of hot water from the kitchen stove, got the water we needed for the house and for the livestock in the barn, then we simply held up the pump handle until the water in the pump drained back into the well. In the kitchen we had a sack of flour and sufficient salt and sugar for a month. We had a closet full of canned fruit and vegetables and as for meat, we literally lived "high on the hog," with crocks stored full of sausage and with hams and shoulders hanging from the rafters of nature's own deep freeze—the woodshed just outside the kitchen door.

We all were kept quite busy caring for the fires and the livestock and in between these chores we caught up on a lot of reading. We resurrected a stack of back issues of the *Saturday Evening Post* and the *Ladies Home Journal* from the storage closet and read the stories we had passed up when the magazines first came. I read *Rolf in the Woods* all over again and this time I particularly enjoyed the chapters that told how Rolf and Quonab had built their cabin and how they had passed their first winter in the snowy north woods. Once again Mother read Whittier's "Snow-Bound" to us all and this time we appreciated, as we never had before, the lines:

> Shut in from all the world without,
> We sat the clean-winged hearth about,
> Content to let the north-wind roar
> In baffled rage at pane and door.

My telescope got a good rest during that snowbound week for the mounting stood out in the front yard half buried in a drift—its grindstone turntable piled high with a peak of white. One of those sub-zero nights was beautifully clear so I bundled up and went out on the back porch. Even my eyes smarted with the cold as I looked about. Jupiter was high above me near the Pleiades. A little further to the east, in Gemini, was Saturn, while just above the barn Mars made a glowing red tip for a lightning rod. Everything seemed to be doing all right without me so I hurried back to my book by the fire.

It happened that during this period of isolation I was expecting some important mail from the AAVSO in the form of my first variable-star charts together with instructions in how to use them in making observations. Day after day I would think how nice and cozy it would be to sit by the fire and arrange and rearrange those charts and study those instructions. Finally I decided to do something about it. Early one afternoon I started on foot for the post office in town four miles away. The roads had now been cleared enough for farm sleds to get through but on foot it was still slow traveling. About a mile from home the road ran past a cemetery and as I approached a strange funeral procession pulled in through the high iron gate. The entire cortege consisted of three bobsleds—the one in the lead serving as the hearse.

I finally reached town and any weariness I may have felt from my long trek vanished the minute I entered the post office and found all the mail I had hoped for waiting for me. In addition to the big manila envelope containing my charts and instructions there had also arrived my first issue of the magazine, *Popular Astronomy*. This Mr. Olcott had recommended as essential to all active AAVSOers, for it published each month the results of their findings.

After tying all the accumulated mail in a bundle for easy carrying I headed back toward home for it was already late in the afternoon. As I walked along in the wake of my lengthening shadow my thoughts went back to another walk to another post office and I remembered how that walk, like this one, had brought me my first copy of a new magazine. Ten years before, in a pre-RFD era, I had walked the half-mile of country road to our combined post office and grocery store to get my first *American Boy.* I recalled how the daily mail train would thunder by and, without slowing down, and with scarcely even a toot of recognition for our four-house crossroads, it would dump off a sack of mail while at the same time a steel bracket would reach out and grab the outgoing sack as it hung suspended from the yardarm right beside the track.

Darkness now was falling as I walked along through the snow and one by one the stars were coming out. I named each one as it appeared: Sirius, the Dog Star, in the southeastern sky; Vega, still in the twilight of the low northwest; high up in the sky was Capella. Then came Rigel and Betelgeuse in Orion, Aldebaran, Procyon, then, low in the east, the twin stars Castor and Pollux, and, finally, I could make out the Big Dipper, balanced on its handle in the low northeast.

By the time I was halfway home it was completely dark. Every star name that I knew had now been fitted to its owner and for another mile or so I tossed a few Greek letters here and there about the sky. The Milky Way arched high above my head and as I was tracing it the tall spruce trees of the cemetery loomed up beside the road and in the gloom I could see where fresh sled tracks now came out the exit gate. Several times an owl in one of the spruce trees inquired who-o-o I was but I had no time to answer him and hurried on toward home.

At one point where the road behind me angled toward the north I stopped and looked back. Vega now was gone but there, low in the sky, the Northern Cross stood boldly upright though its foot was slowly sinking in the snow.

Supper was waiting for me when I finally got home. There was mail for everyone and the lamp and the fire burned late that night.

9 Variable Stars and Tropic Isles

LIFE WAS NEVER QUITE THE SAME FOR ME AFTER THAT WINTER walk to town. The charts that I brought home with me were potent and ensnaring and I feel it my duty to warn any others who may show signs of star susceptibility that they approach the observing of variable stars with the utmost caution. It is easy to become an addict and, as usual, the longer the indulgence is continued the more difficult it becomes to make a clean break and go back to a normal life. As this is written, forty-six Januaries after that snowy jaunt, I have set my alarm clock for 4 A.M. three times thus far this month and gone out through the snow to my observatories just to spend a chilly hour or two with the predawn variables.

Variable stars came my way at a most propitious time. My telescope and I had now explored the skies for many months. Like some celestial Captain Cook we had been slowly sailing through a far-flung shoal of deep-sky islands. But now, guided by my charts, there loomed on my horizon something quite unlike those calm unchanging isles of cluster, nebula, and double star. A variable star was a completely new experience; it was not just something that was THERE, it was something that was HAPPENING!

From the little instruction manual that came with my trial set of charts from the AAVSO I learned that a variable star is simply a star whose light output is not constant. At certain times such a star may be hundreds or even thousands of times brighter than at other times. As an example, on the night I walked home from the post office carrying my charts and instructions I saw the Northern Cross setting in the northwestern sky. Now, had I been familiar with all the information which, at that moment, was tucked securely under my arm, I would have taken a more careful look at that well-known figure. Were there, that night, four stars or were there five in the long axis of that cross that began with bright Deneb at the top and ended with yellow Albireo at the foot? If there were five stars then the one just above Albireo was the famous variable star, Chi Cygni. This star is visible to the naked eye for about six weeks out of every fourteen months. When at its faintest a 4-inch telescope is required to see it. Variable stars, like those who watch them, have their ups and downs.

Variety is a universal spice. If human beings all looked alike, behaved alike, and thought alike life could get very monotonous and the same is true of variable stars for if all variables were precisely like Chi we certainly would not lose a great deal of sleep over them. Actually Chi is just one member of a large family—the Long-Period Family—so named because they require two hundred to four hundred days to go through a complete cycle of variation in brightness. Chi has many brother and sister variables and they all have their own little individual characteristics but there is no mistaking them, for each one has that slow, deliberate manner that is typical of the family. Chi has a lot of cousins, too, ranging from first and second cousins like Semi-Regular and Red Giant, all the way to forty-second cousins like the Cepheids. These Cepheids are a loquacious lot who have tattled many secrets of the skies, but it doesn't take the astronomers long to find out just how bright they really are. Then they put them in their proper place. Also distantly related are the members of the Cataclysmic Family. There are some dwarfish albinos in this line. As a family they will bear close watching for their behavior is quite unpredictable and highly irregular. No blood relation at all is the Eclipsing Family. The members of this clan are all very precise and dependable but so utterly futile. They just go around getting in each other's way!

In actual practice I found that observing variables is quite simple. It consists merely in comparing the brightness of the variable with the brightness of other stars in the same field of view which are not variable and which have had their magnitudes accurately determined and indicated on the chart. My only problem was in locating the variable.

First of all I had to have a star atlas so that I could plot, on its maps, just where the variable was located in the sky. A star atlas is actually a geography of the sky. It is usually made up of six or more maps which show all the naked-eye stars over the entire area of sky. Like earth maps with their meridians of longitude and parallels of latitude, sky maps are crossed by hour circles and declination circles. The purpose of all such circles is to locate positions accurately. Thus my home town of Delphos is located at 84 degrees west longitude and 41 degrees north latitude while, in the sky, the bright star Vega is located near 18 hours of right ascension and 40 degrees north declination. The Upton's *Star Atlas* which I purchased that year served me well until it was completely worn out thirty years later. It had two features that would make it somewhat outdated today. It still carried the old mythological figures of the constellations and it, of course, also had the old irregular constellation boundaries. These boundaries were revised in 1930 by the International Astronomical Union so that they now coincide with arcs of right ascension and declination.

As soon as my star atlas arrived I got out my charts and, working very carefully, I made a tiny dot and circle on the maps in the precise location of each one of my variables. It was now late February and the skies, for seemingly endless nights, were completely clouded. Finally, came a change of weather and with it a clear, cold night. As soon as darkness fell I bundled up and, with telescope, atlas, and charts in my mittened hands, I went out to find my variables. Two hours later when I returned, half frozen, to the fire I had not found a single one.

Nor was I any more successful on several succeeding nights and I was becoming more and more discouraged. Finally, however, on the night of March 1, 1918, I set the telescope up near the northeast corner of the house to keep out of the wind and got out my chart of the variable, R Leonis. I pointed the telescope at the fourth-magnitude star Omicron Leonis and, using it as a base of operations I started exploring the adjacent territory. To my great delight, about a minute later and just a little more than one field to the northeast of Omicron I found the tiny triangle of stars with R, my first variable, forming one angle precisely as shown on the chart.

Why had success been so long in coming? Simply because I had no mental image of what to look for. I had not bothered to figure out just how much of the chart area was covered by the instrument's field of view. After this experience I soon made a wire ring which could be moved about over my charts to show me just what my telescope was seeing. In the nights that followed other variables were located with comparative ease, but every March first since that night, whenever the skies were clear, R Leonis and I have recalled our first meeting by making a mutual estimate of our brightness.

The single estimate that I made that night was the beginning of a long accumulation of more than a hundred thousand other estimates to follow. And the very meager monthly report that I sent in to the AAVSO at the end of that March in 1918 began an unbroken sequence of 552 consecutive monthly reports to the present date—46 years later. Verily, one can easily become an addict.

So many of the variable stars are total nonconformists and, for me, this constitutes their greatest charm. It is their unpredictable behavior, more than any other factor, which has so long sustained my interest and has made the watching of them a literal *ten* "Thousand and One Nights" of entertainment.

When I go out tonight and train my glasses on that sky-wide stage I am quite certain to find that, among my hand-picked cast of erratic actors, some have thrown away the script and are now extemporizing. It may be that R Coronae, after years of shining as a naked-eye star, has tonight begun a fade-out that will only end when it is near the limit of my largest lens. Or perhaps U Geminorum, roused from months of faintness, suddenly has brightened overnight. As I turn my scope to the north I wonder just what Z Camelopardalis and TZ Persei may be doing. One never knows from night to night for their behavior is completely baffling. I have seen them apparently get "stuck" about halfway between their brightest and faintest limits and remain there, without noticeable change, for ten months at a time. V Sagittae, a novalike variable, usually is putting on a show, while its classmate, Z Andromedae, is only waiting quietly to catch me off my guard. I must not forget tonight to glance at Gamma Cassiopeiae, the middle star in the familiar "W." For many years now it has not changed, but it can, for in 1937 I watched it brighten until it nearly equalled first-magnitude Deneb in the Northern Cross!

There is nothing in all the world of the sky that will provide a more lasting source of interest than will these amazing antics of the variable stars. And, as I learned many years ago, a 3-inch telescope is not required to feed this interest. It can, in fact, be made to flourish with no telescope at all.

So many times, while showing the stars to others, I have heard this plaint: "I would like to watch the stars but I have no telescope." So many letters have asked: "Where can I buy a telescope so that I can study the stars?" Stars and telescopes have come to be regarded as inseparable. This is most unfortunate. I may frequently refer in these pages to the superlative delights of a telescope and to the enchantments of its deep-sky probings, but I would also like to state most emphatically that a telescope is not essential to an enjoyment of the stars; that even without optical aid of any kind one still can become an accomplished star-gazer. No one, as yet, has ever nearly exhausted all the possibilities of observing with the naked eye alone.

Sometimes, when I am watching through my telescopes, the question comes to me—what would life be like without a telescope? How would I use the time that now is devoted to observing? Always the question leads, by a long and devious train of thought that takes me half around the world, to the conclusion that life for me would change but little. Telescope or not, I would still keep watch.

Ever since my early boyhood some of my most avid reading has been concerned with shipwrecks and castaways. *Enoch Arden, Swiss Family Robinson, Robinson Crusoe,* and Jules Verne's *Mysterious Island* have all held me enthralled for hours on end. They still do, and I have often imagined myself in the role of the sole survivor, waterlogged and weary, staggering ashore on some deserted isle, without a chart, without a compass and—*without a telescope!*

I have noticed that most of these castaways were able to salvage sufficient plunder from their sinking ships to set themselves up in pretty cozy style on shore. In fact some of them never had it so good back home. However in my own Saga of the Stranded Star-gazer I do not require all this bounty from the briny deep. I only ask that I escape from Neptune's wrath with some paper and a couple of pencils!

First of all, while I still remember the date and the day of the week on which I landed, I start my calendar to working. My fellow castaway, Crusoe, kept his lonely log by cutting notches in a post, making an extra long notch for Sundays and a double length notch for the first day of each month. Fortunately I still recall the old reminder—"Thirty days hath September," etc., so I shall do likewise. Crusoe occasionally forgot to make his daily mark and so, perhaps, shall I but it should be possible to make a semi-annual check on my calendar provided I am favored with clear horizons at the times of the solstices in June and December. By means of very careful sightings I should be able to determine the date when the sun rises and sets furthest north of the equator about June 20, as well as the extreme south points reached on December 20. Correct dates are of the utmost importance in any kind of record keeping, whether in Ohio or its antipodes and any observed phenomena, such as eclipses of the sun and moon or occultations of bright stars by the moon will, if recorded, serve as later checks on the accuracy of my island calendar.

Without a doubt the first question that comes to the mind of every shipwrecked survivor is: "How do I get out of here?" Sometimes a little sober reflection might change that "how" to a "why." After all, Enoch Arden went home only to find his wife, Annie Lee, happily married to his early rival. The crew from *The Mysterious Island* ended up as six bachelors on a farm in Iowa. In 1898, when Captain Joshua Slocum, while on the first single-handed voyage around the world, put in at the island of Juan Fernandez, he visited the cave where Alexander Selkirk, the real-life counterpart of Robinson Crusoe, had lived for more than four years in complete solitude. Slocum's parting words were, as he sailed away: "Blessed island of Juan Fernandez! Why Alexander Selkirk ever left you was more than I could make out."

Perhaps the next concern of the castaway will be the question: "Where am I?" On the *Mysterious Island* Captain Harding immediately set out to determine their longitude and latitude. He figured their longitude in what always seemed to me a rather sneaky way—his watch was still running on Eastern standard time! To utilize this invaluable asset Harding drove a stake in the ground and then carefully noted the time by his watch when the shadow cast by the stake was at its shortest. This occurred at their 12 noon on the island but by his watch it was 5 P.M. It was, therefore, 5 hours later in Washington, D.C. than on the island and, as the earth turns on its axis 15 degrees per hour, this meant that their island was located 75 degrees to the west of their home meridian, or 150 degrees west of the Greenwich or prime meridian.

Since I made no such timely arrival on my island the question of longitude is not for me. Of far greater importance to me is the matter of latitude, which has a much more direct bearing on temperature and climate. A single look at the night skies would suffice to show my latitude with reasonable accuracy, but a more precise determination can easily be made simply by measuring the angular height of the pole above the horizon. There is no conspicuous star near the south pole of the heavens but a few night sightings should soon locate the stationary spot in the sky around which all the neighboring stars revolve. Having located the pole I take one of the sheets of paper that I brought ashore and, holding it firmly against some vertical support, I sight along its long lower edge to the distant sea horizon directly beneath the pole. Without moving either my eye or the paper I next sight toward the polar point and mark this location on the paper. I then draw a straight line between this polar mark and my eye location on the paper. My latitude is equivalent to the angle included between the line that I have drawn and the bottom edge of the paper.

It is only necessary now to measure this angle in degrees to know just how many degrees my island lies north or south of the earth's equator. There are several ways I can divide the angle I have drawn into the proper number of degrees. I could use the moon's apparent diameter of one-half degree as a convenient unit or, since the constellation Orion is visible, in season, from every desirable deserted island on earth, I could use the three-degree length of his belt as a still larger unit. By still another method I could point my paper angle at some star near the celestial equator, such as Delta in Orion or Eta in Aquila and, with the paper lying in the plane of the star's travel across the sky, I could mark the angle traversed by the star in four minutes' time. This will be just one degree. I measure this four-minute interval by my pulse, which has a most convenient rate of 60 beats per minute. Therefore 240 beats equal one degree. Come to think of it, this is just about as sneaky as Captain Harding's watch!

I have never ceased to marvel at the almost incredible feats performed by some of the early astronomers long before the invention of the telescope. In 250 B.C., without any instrumental aid, Eratosthenes measured the distance around the earth—and he never left home! He then measured the inclination of the ecliptic to the equator and, at the same time, made the first latitude determination in history—that of his home town of Alexandria, Egypt, and with an error of less than ten miles. Another early great of that same period was Hipparchus who noticed the slight wobbling movement of the earth which shows up in the twenty-six-thousand-year shift in the poles and in the tiny annual precession of the equinoxes. Hipparchus also made the first really good chart of the heavens, being inspired to do so by the sudden appearance, in the year 134 B.C., of a bright new star. With this chart he prepared himself to recognize the nature of any future strangers in the sky.

As a boy, still in my pre-spyglass years, I had been fascinated by a full-page picture, in our Ridpath's History, of Hipparchus standing before his open observatory in Alexandria, measuring angular distances in his ancient sky with an instrument known as a cross-staff. I, too, soon fashioned a cross-staff that used the 5 degrees between the Pointer stars of the Dipper as a unit of measurement and with it I happily figured angles from star to star across the skies of our Ohio farm. Now, once again, I follow in the footsteps of Hipparchus as I construct a star chart while on my lonely isle.

I draw with particular care the entire Milky Way region so that the location of every star that I can see is shown just as accurately as I can make it. My reason for all this painstaking detail is this: I now have the abundance of time that is necessary for a really thorough and systematic search for new stars and, on the average, I should succeed in finding one every three or four years. Often in the past new stars bright enough to have been seen with the naked eye have been found later on photographic plates and reported to Harvard College Observatory—the focal point for all astronomical telegrams. With persistent watching of the skies I should be able to catch some of these new stars—perhaps even on their opening night.

There also will be quite a number of my old friends, the variable stars, who will keep me company. R Leonis, R Hydra, Omicron Ceti, R Serpentis, Chi Cygni, and the irregulars, R Coronae and R Scuti, all, at times, are visible to the naked eye and all will visit me from time to time. Some of them—perhaps all of them—may be quite invisible when I first start watching from my sea-bound station but nevertheless, each starry night I will search their empty, blank locations in the sky until finally the night will come when the first faint flicker of a gleam appears. Night after night, week after week it will brighten, then, just as slowly, it will fade until at last its place is blank and empty once again.

Still other blank spots in the sky I shall also watch each night—spots where once-bright novae now are sleeping. Sometimes these slumbering giants have come to life again! Each night too will have its streaking meteors and occasionally a fireball will be seen and charted. I can always hope for another meteor shower such as I saw on the evening of October 9, 1946, when the earth passed through the swarming fragments trailing in the wake of Giacobini's Comet which had passed that point of intersection with our orbit only a few days before. For half an hour on that night the sky seemed full of meteors, all streaking from the low northwest. At the height of the display one meteor per second could be counted—some of them as bright as Venus!

The Gegenschein—a rarity reserved for perfect moonless nights—I shall also see. This phenomenon, which is only faintly seen at best, appears as a circular, hazy spot a few degrees in diameter, exactly opposite the sun's location in its path around the sky. It is best seen in the month of October when it may be found in the faint constellation, Pisces. It is thought to be caused by the sun shining on tiny meteoritic particles that surround the earth. But no one knows for certain.

I shall do some comet hunting too, and mainly my attention will be focused on the region close about the sun. With the sun's disk shielded I shall often do my searching while the sun is in the sky. Many comets of the past—and this includes my childhood friend of early 1910—first appeared as daylight comets.

The auroras—lifelong favorites of mine—I shall greatly miss. For, ever since I was exposed, while in my callow youth, to Gilda Gray's gyrations and the enraptured reading of O'Brien's South Sea tales, my isle must be replete with swaying palm trees. And coconuts and auroras just don't go together.

One thing, though, bothers me. Supposing I do find a new star or a comet while on my tropic isle—where will I find a bottle to carry my message to Harvard?

10 School Days

April brought the last of my school days. A war was on; a war that would end all wars and make the world safe for democracy. My older brother Kenneth was with a unit of the heavy artillery somewhere in France and, as the spring work was starting on the farm, it was thought best that I replace him at home for the duration and then go back later and finish high school. To this I voiced no objection. I was young and youth has always welcomed change.

I had enjoyed high school tremendously. It had been a whole world different from the monotony of the country school of my early years. It had brought new scenes, new acquaintances, and new interests. Not once while in high school did I try to escape attendance with a feigned sore throat.

With high school came an entirely new category of studies. Here I became aware of English literature and was formally introduced to such immortals as Shakespeare, Tennyson, and Sir Walter Scott. Their complete works had reposed in our bookcase on the farm all my life but I had never read them. This was most fortunate. To have met them earlier than I did would only have thrown out of line the smoothly meshing wheels of time and circumstance. Had I, at twelve, listened to the moonlight tryst of Lorenzo and Jessica it would have been but wordy prose. At sixteen it was melody. When I read Scott's "Lady of the Lake" the skirl of bagpipes was echoed by my fife, and I saw the moon that danced "on Monan's rill" reflected in the Auglaize River as I leaned on the railing of the old iron bridge. Locksley Hall came to me at a time when I too was nourished "with the fairy tales of science." At a time when I, like Tennyson, also watched the stars through a 2-inch telescope.

Biology and botany were my favorite subjects while in high school. I had a year of each of these studies and in looking back I cannot now recall any specific thing that I may have learned on either subject while in the classroom. I may have chanced to wander through those ivied halls just at an unfortunate time but, as I remember it, very little was done to promote any lasting affinity between the student and those particular subjects. At that time visual education seems not to have been invented and field trips were frowned upon. Two good Spencer microscopes reposed under bell jars on the same shelf beside the spyglass that I had longed to look through just the year before. I never got to use them.

In botany we never once tried to identify a plant by using the floral key in our textbook. In fact, the botany teacher seldom brought up the subject of botany at all! He had come to us fresh from Johns Hopkins University and, unfortunately for us, he brought his textbooks with him. We took copious notes on his lofty lectures from E. B. Wilson's classic work, *The Cell.* I still find in my old notebook page after page devoted to histology, microtome technique, and the staining and preparation of bacteriological specimens.

In spite of all this I enjoyed these two studies for both the biology and botany textbooks made fascinating reading and they also served a very solid purpose for they became my introduction to the nomenclature and classification of the plant and animal kingdoms. Ever since the days of Rolf I had been collecting, mounting, preserving, and transplanting everything from the surrounding fields and woods that didn't actually fight back and now these assorted acquisitions took on the added dignity of species, genera, and family.

One unexpected bonanza came from my perusal of the biology text. I discovered a new world! It was the unseen world of the hay infusion. Following the suggestion in the book I stuffed a handful of timothy hay from our mow into a quart mason jar and then filled the jar with rain water. After this mixture stood for two days in the warm sunshine of my bedroom window a thin scum covered the surface of the water. Transferring a drop of this liquid to a small fragment of window glass, I maneuvered it into the focus of the high-power eyepiece from my spyglass. To my amazement that cloudy drop was inhabited by a myriad of tiny living creatures all scurrying about their little sphere on circumnavigating voyages. Fascinated, I followed their great-circle courses with my lens until at last they all had run aground on the silicon shores of their drying sea. Again and again I made new global seas through which plowed new armadas of Paramecium and Colpidium until their undulating oars of cilia slowed to a final stop.

I tried to envision in my mind what an infinite galaxy of worlds our haymow held in bonds of arid dormancy. Surely they must outnumber the countless stars of the sky. I wondered, too, how many spyglass lenses had ever watched two such extremes—the microcosmic worlds in water drops and the giant orbs of the Milky Way.

My only high school problem was one that I had in common with all rural students then—the problem of getting there. At that time there was no convenient school bus, with its pickup and delivery service, such as we have today. We had to get there under our own power. In my case this was usually by bicycle and, as school was four miles away, this means of transportation was far from pleasant in unfavorable weather. No doubt this daily inconvenience helped me to accept my dropping out of school in such a placid manner, but the main reason that I welcomed the change in my affairs was the fact that, at that time, winning the war was uppermost in the minds of everyone and farming seemed to be much more important than acquiring an education.

A higher education was not then regarded as the indispensable adjunct that it is today. The country grade school that I had attended consisted of just one large room. Here one teacher taught all eight grades in such varied subjects as reading, writing, spelling, arithmetic, history, grammar, physiology, and geography. This meant that the teacher had about twenty classes to listen to every day—five of these in reading alone—while at the same time he attempted to preserve a semblance of decorum in the rest of the assembly. In his idle moments he fed the big, hungry coal stove which stood in the center of the room.

A blackboard ran the full length of the long north wall and above this hung three large pictures. On the right was a portrait of a stern and dignified President McKinley seated between his wife and mother. In the center picture the battle of Mobile Bay raged at full speed regardless of torpedoes, while the picture on the left was an arctic scene showing Eskimos, igloos, polar bears, and icebergs. This picture may have had a title but I have forgotten it.

In one corner of the schoolroom sat an ornate but wheezy foot-powered parlor organ which two of the older girls alternated in playing for the song session that started off each day's quest for knowledge. We pupils were always permitted to choose the songs for these musical moments and, quite unaccountably, the outstanding favorite was the song entitled, "Work for the Night Is Coming." Other frequently requested numbers were "Juanita," "Swanee River," and "Ding Dong Dell." We sang all these in unison for we knew nothing of part singing—which may well have been a blessing in disguise.

I recall one incident that took place in one of my classes during my first year in school, an incident which, in view of my later concern with the doings of the universe, seems something less than auspicious. There were only three of us in the class that day—two little girls and myself. We were probably studying the First Reader, but of this I am not certain. I only know that it was an ultra-elementary class of some kind and that the teacher had asked the question—"What is the shape of the world?"

"It's round," replied one little girl.

The second girl added her bit, "It's like a ball."

"That's right," replied the teacher. "Now then, where do we live? Are we inside or outside this ball?"

Both little girls in unison, "Outside."

Little boy's reply—drowned out by girls, "Inside."

I waited expectantly for the teacher to set the girls straight after their silly answer but it was recess time and the class was dismissed without further comment. I pondered over these two opposing theories the rest of the afternoon and I took the problem home with me that night where my mother failed me completely by favoring the girls' side of the affair. But I didn't give up without a struggle, for in the blue vault of sky by day and in the bowl of stars by night I could plainly see the inner surface of my world. O well! Greater minds than mine have seen the world as flat or even as supported by four elephants standing on a waddling turtle!

Each Friday afternoon we closed the week's activities with an hour or two of sheer frivolity. Actually it was a sort of postman's holiday for we were given our choice of either "spelling down" or "ciphering." The first of these was, of course, a spelling contest. The teacher would appoint two opposing captains who would "choose up sides," each captain alternately choosing a pupil until we all were ranged along the opposite sides of the room. The teacher would then pronounce the words and whenever a pupil missed a word he sat down and the opposing side tried it. Ciphering was conducted along similar lines, though here it was a race against time in solving problems in simple arithmetic at the blackboard. Sometimes, in the mild weather of spring and fall, these Friday afternoons would be given over to a baseball game with some nearby district school. Our team was noteworthy in at least two respects. We had the only girl player in the league. Addis Ludwig was the tallest member of the team and played first base with the best of them. Also, two of our nine players (my brother Kenneth and Myron Foust), later on became semipro players.

We were given a fifteen minute recess at ten o'clock each morning and another in midafternoon. Our noon hour started off with a dash to the southeast corner of the schoolroom where all our lunch boxes were ranged on two long shelves. Ten minutes later the lunches had disappeared and the room was deserted. Outside, on the acre of schoolground, several games would be forming; games of baseball—if anyone had a ball and bat—or of "prisoner's base" or "blackman," both of which required no other equipment than two good legs. Another group—the temporarily affluent—would walk the dusty quarter mile to Tobe Luttrell's grocery store for a penny's worth of candy corn or a stick of the round, white, pencil-size "Long Tom" chewing gum which later that day would finish its flavorful career on the underside of a desk or seat or, if tossed outside, would eventually gain some added mileage through a pickup by a passing shoe sole.

One afternoon at school we had a special recess. It was on this occasion that I saw my first eclipse of the sun. Mr. Humphreys, our teacher, knowing of the impending affair, had brought along that morning a number of small pieces of glass. At noon he removed one of our six oil lamps from its bracket on the wall and placed it on a desk. When it was lighted we held the glasses just above the flame until a small area near the end acquired a heavy black coating of carbon. At the time scheduled for the eclipse to begin we all went outside and trained our smoky glasses on the sun. The eclipse already had begun for we could easily see a small black notch in the sun's western edge. We watched for about fifteen minutes then, as the eclipse would be only a partial one at best, we went back inside where Mr. Humphreys drew a large diagram on the blackboard and explained to us the mechanics involved in what we had just seen.

After school was dismissed for the day I lingered behind a little and told Mr. Humphreys that I had seen a little black speck on the sun while we were watching the eclipse. He had not seen it but said that undoubtedly it was a sunspot. When I asked what a sunspot was he replied that it was thought to be a hole in the outer surface of the sun. After the schoolhouse was locked up for the night we two walked homeward together until our roads separated half a mile to the north. As we went along he talked of a trip that he and his brother had made that summer to visit their family and relatives back in Wales. We were approaching the crossroads when I recalled something—probably brought on by the events of the afternoon.

"Mr. Humphreys, do you know the proper order of the three outer planets?"

"No," he graciously replied, "I'm always getting them mixed up."

"They are Saturn, Uranus, and Neptune," I expounded, "and their initials spell SUN."

I had only recently acquired this little gem from my mother, and to be able to pass it on to a schoolteacher made it all quite a day for me.

Mr. Humphreys was my teacher for four of my eight years in the country school. He had progressive ideas far in advance of the times. For all such holidays as Thanksgiving, Christmas, Washington's Birthday, Arbor Day, and the last day of school he arranged programs in which each pupil had a part. He tried to teach us citizenship along with our textbook studies. We even held court occasionally to try our own misdemeanor cases. A day or two before each state and national election we held our own elections. Most of these school elections were Democratic landslides but we managed to elect Taft in November 1908, and Mr. Humphreys was visibly elated. Tom was a dedicated teacher and a kind and sympathetic friend.

The grading system in these country schools was a rather vague affair. One grade just sort of blended into the next and, in general, school was something that one eventually outgrew rather than graduated from. Many of the older pupils were eighteen or twenty years old before they finally left for other endeavors. This is in no way a reflection on the intelligence of those students; there simply was no incentive for a higher education unless the student intended to become a schoolteacher rather than a farmer. Even the school term itself was often a matter of personal convenience, for it was customary for the older boys to start their school term after the cornhusking was finished in the fall and these same ones left each spring as soon as the ground was dry enough to start plowing. To my knowledge, I was the first alumnus of that country school to go on to high school and even then, as has already been related, I became a dropout victim of those wartime years.

The tremendous wave of patriotic fervor engendered by World War I here in my home town was something never experienced before or since. We were waging a war against Militarism and we were assured that with victory would dawn a millennium of peace on earth. No personal sacrifice was too great if it would further the war effort in any way. Our various fund drives were all oversubscribed; we had meatless days, wheatless days, and sweetless days. We ate war bread and we sweetened our coffee with corn syrup and we liked it!

Here and there, of course, some went much too far in their patriotic zeal and did things which were just plain silly. Our town had been founded by emigrants from Germany and throughout its history the greater proportion of its citizens have been of German descent and many of these families still spoke that language in their homes. In the front windows of these same homes one could usually see veritable constellations of service stars, for these families, as a rule, were large. But it was a season of distrust and here was a fertile soil for germinating the seeds of suspicion. On several occasions these sprouted into ugly growth and self-appointed groups made up of some of the town's respected citizens caught the vigilante virus and forced their bewildered neighbors to publicly kiss the flag. Freedom of speech was a dangerous thing during World War I. In my final month of high school our class in German grammar was discontinued by order of the Board of Education.

That summer, for the first time, the farmer prospered in a modest way. A bushel of wheat brought $2.20. Clover seed rose to an all-time high of $30.00 per bushel. Land was at a premium and sold for as much as $300.00 an acre. Strawberries, our longtime specialty, which in my spyglass days brought a top price of 15 cents per quart, rose to an unheard-of price of 60 cents that June. But that was also the first June that we had none for sale. Dad had said: "We can hit the Kaiser harder with an ear of corn!"

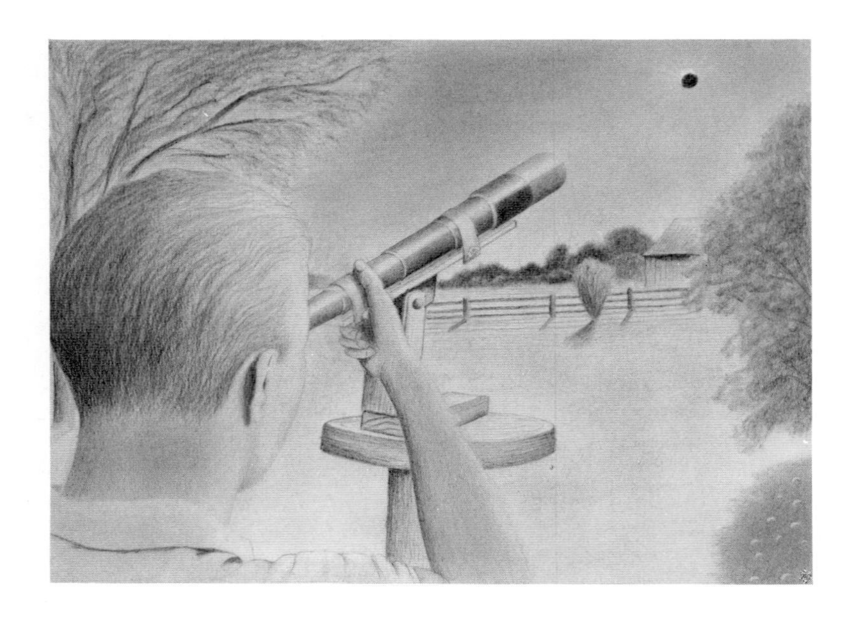

11 June Spectacular

JUNE 8, 1918 IS A DATE I SHALL NEVER FORGET. ON THIS DAY, IN late afternoon, the shadow of the moon would march across America. Once again I would see but a partial eclipse, in most respects a duplicate of the one I had watched through a smoked glass from the country schoolyard just three fields and a woods away. But that was long ago and I knew that I had missed many details of that first event.

For the past several months I had been reading about the

coming spectacle. It would be a total eclipse in a narrow belt, not more than sixty miles wide, running diagonally all the way across the country from the State of Washington to Florida. On either side of this thin line would be another belt two thousand miles wide in which the eclipse would be only partial. Here in Ohio, more than five hundred miles north of the line of totality the sun would be about 75 per cent covered by the moon.

It would be the first total eclipse of the twentieth century in the United States and in spite of wartime restrictions and in spite of the fact that many astronomers were in active service—some at the front, others doing computing and teaching navigation—nevertheless it promised to be the most completely observed eclipse in history. Most of the eclipse expeditions were setting up their camps along the western end of the totality path—for a number of reasons. There the duration of totality would be at its greatest. In Washington and Oregon the sun would be covered by the moon's disk for a full two minutes; in Oklahoma for a minute and a half; while in Florida it would last slightly under one minute. But totality duration is only of secondary importance compared to that always uncertain atmospheric factor—clouds. Many an eclipse expedition has traveled halfway around the world, spent weeks in setting up elaborate equipment and rehearsing its carefully planned program, only to have a cloud drift across the sun just before totality. A site is not selected until its shady past is thoroughly investigated and its cloudy-sky percentages for that particular day and hour in former years are carefully weighed and compared with those of all other possible sites. Here again, the more arid West was heavily favored. One final advantage of the West was that the eclipse would begin there a full forty-five minutes before it would arrive in the East and the sun would therefore be higher in the sky and less affected by the heavy atmosphere of the horizon.

Even though some certain locality along the path seems to offer superlative advantages over all others it would be the height of folly for all parties to settle there and gamble everything on the whim of some tiny cloud. Whenever it is at all possible the various camps are strung out along the path so that some of them, at least, may be successful. The party from Lick Observatory located at Goldendale, Washington. At Baker, Oregon, was the Naval Observatory camp, while both Yerkes and Mt. Wilson decided on Green River, Wyoming, for their stations. In addition to these a large number of smaller parties located in central and eastern Colorado.

In reading up on eclipses in preparation for the coming event I learned that from two to five solar eclipses visit the surface of our planet each year. However, some of these may be only partial while others, though total, may strike the earth in inaccessible regions such as the poles or their line of totality may fall entirely on the vast surface of the sea. On the average the astronomers can expect a good observable total eclipse of the sun about every third year. I learned, too, that any particular spot on the earth might look forward to such an eclipse once every 360 years. This too is just an average for some localities seem to be much more fortunate than others. Denver, Colorado, for example lay right in line with the coming June 8 blackout and only forty years before, in 1878, it had also been in the path of a previous totality. At the other extreme I could find no record that our farm had enjoyed any such solar spectacle within historic times and certainly there will be none within the present century.

My mother thought that she remembered seeing one as a small child, but if so, it could not have been from this region. She did, however, recall in detail, as a relic of her schoolteaching days, a poem by Whittier entitled "Abraham Davenport." This poem vividly describes the celebrated "Dark Day of New England," a day so dark that:

> Birds ceased to sing, and all the barnyard fowls
> Roosted; the cattle at the pasture bars
> Lowed and looked homeward.

It was a day of such weird gloom that:

> All ears grew sharp to hear the doom-blast of the trumpet
> Shatter the black sky.

From all the phenomena cited in the poem my mother quite naturally assumed that it referred to a total eclipse of the sun, but in delving further into the records—particularly those furnished me by the Ferguson Library of Stamford, Connecticut, the locale of the poem, I find that the reference was simply to an extremely dark midday cloud overcast which may have held an added mixture of fright, superstition, and smoke from forest fires, for there was no eclipse of any kind in New England on May 19, 1780, the date recorded in Whittier's poem.

Actually there was a total eclipse of the sun in New England that same year but it did not occur until October 21, a full five months after the "Dark Day," and it was total only in the state of Maine. This eclipse was noteworthy in that it marked the first organized American eclipse expedition. This occurred during the Revolutionary War and at that particular time the British forces occupied that part of Maine wherein the path of totality lay. However, the British garrison at Penobscot generously granted permission for an eclipse party from the Commonwealth of Massachusetts to choose an eclipse site in that vicinity. This eclipse marked a truly "dark day" for American astronomy for, due to either some miscalculations or to faulty moon tables, the expedition failed to quite locate in the totality belt and consequently saw it only in its partial phase.

On my long-awaited morning of the eighth the sun rose clear and bright and throughout the early hours of that rare June day I watched the sky as anxiously as did any of the astronomers along the central path. As I drove the cows to pasture I noticed that they made long dark trails through the sparkling sheen of the dew-drenched grass. This I had always found a good fair-weather sign. About ten o'clock a sudden shadow, falling on the field of inch-high corn that I was cultivating, made me stop and look up quickly. High, fleecy clouds were riding eastward in the deep blue sky. Long a confirmed cloud watcher, I singled out a little wisp of white and carefully surveyed it. Right before my eyes I saw it shrink and disappear, but it left me with another omen of clear skies.

About three o'clock that afternoon I rode the clanking cultivator back to the barn and put the sweaty team away. It was time to get my eclipse station ready for the big event. First of all I called Western Union on the telephone, got the correct time, and brought our battered Ingersoll alarm clock up to date. Then I clamped my spyglass on its grindstone mount and carried it out to the front yard where, in the shade of one of Grandpa's twice-planted walnut trees, I had a good clear view of the southwestern sky. The morning clouds which had speckled the farm with shadows had long since disappeared and with them went my weather worries. I was certain now that at least one station would be favored with fair skies.

On my June afternoon I knew full well that a partial eclipse had no scientific value and that no respectable astronomer would be caught looking at one, but it was my first opportunity to watch one with anything more formidable than a piece of smoked glass. Now I had a telescope and that telescope had a deep-red sun filter as an accessory for its high-power eyepiece. I intended to make the most of it.

A glance at the alarm clock told me that in just two minutes the warning tocsin of first contact would bring the deepening shadow of the moon out of the Pacific and start it on its swift cross-country trip. Even so, that flying shadow would not reach the farm for more than half an hour. So, with everything in readiness and some time to spare, I turned the 2-inch on the sun.

The year 1918 was near a sunspot maximum and through my filtered lens the red-faced sun revealed a scattered rash of pockmarks. Seeing them thus recalled to me my first eclipse ten years or more before when, at the previous maximum, I had found a giant sunspot—without a telescope. Several times during 1916 and 1917 I had seen other naked-eye sunspots; once on an early morning when the sun shone dimly through a clearing bank of fog; again on an autumn evening as it set in a drifting pall of smoke from far-off forest fires. But most of the giant spots that I had seen—each one at least four times larger than the earth—had announced their presence on the previous night by a vivid fanfare of auroral streamers in the northern sky.

I watched the sun for a long time that day as I waited for the eastward speeding shadow to overtake the eastward spinning farm. It was the longest session that the sun and I had ever held and I found it hard to realize that I was looking at a star. Then my mind became involved in wondering why an astronomer would spend the long dark hours of the night in worrying about some tiny star a billion light years distant at the very limit of his sight, then sleep throughout the day while the nearest, brightest star of all was shining just outside his darkened window.

Just then the alarm clock sounded. The shadow should be coming quickly now; it might, right then, be crossing Indiana. I shifted the telescope ever so slightly toward the western edge of the sun—then, a moment later, I was staring spellbound as the moon, right on time, took its first little nibble from the red-hot cookie of the sun. Slowly, inexorably, the moon moved eastward as it ate its way into the sun with the nicked and broken teeth of its mountained leading edge. Sunspot after sunspot was swallowed by the black invader until at mid-eclipse they all had disappeared, leaving nothing but a crimson-colored crescent sun with downward pointing cusps. Then I recalled that weird outrage recorded by another spyglass wielder—"The Ancient Mariner"—who saw "The hornéd moon, with one bright star within the nether tip." What I now saw seemed stranger still—a hornéd sun with one dark moon within its nether tip.

At mid-eclipse I turned away and looked about. Everything I saw, the nearby fields, the distant vistas, all seemed wrapped in some unearthly early twilight. The sky seemed darker—shadows faint and indistinct. A cool wind, almost chilly, had sprung up from the west. The grass beneath the nearby maple now was appliquéd with scores of crescent suns, projected there from each small aperture between the leaves above.

Back again at the telescope I could see that now the darkest phase had passed. Seated atop my low stepladder I watched, fascinated, as the moon, now in full retreat, slowly relinquished all the solar spoils which it had won. From behind the low serrations of the profiled mountains of the moon, one by one the sunspots now emerged from occultation. When finally the last black segment of the moon had dwindled and disappeared I realized that I had just been witness to a strange event that takes place unseen every month; a brand new moon had been transfigured from the body of the old. Only during an eclipse of the sun can we note the instant when the old moon, moving eastward, crosses the median line of the sun and becomes a fresh new moon just starting out on another monthly lifetime.

All over America the eclipse was ended. "The moving finger writes and, having writ, moves on." Like a moving finger of darkness the cone-shaped shadow of the moon had dipped down, scrawled its brief two-minute mark of night across the land and then moved on, still writing, but now with invisible ink upon the empty page of space.

Along the narrow track of totality astronomers from all over the world packed up their precious plates and prepared to leave for home. Weeks before, they had assembled here and had carefuly taken their places in line in order to see a spectacle that would last just two brief minutes. For the most part they left well pleased with the performance though, as always, some had been unfortunate in their choice of seats along the lengthy aisle. And as they started homeward not one in all that far-flung audience could know that this was just an intermission and that the show they came so far to see would be a double feature.

When darkness came that evening I clamped my spyglass to the grindstone mount which still was standing at the station underneath the walnut tree. I hoisted it up on my shoulder and carried it out to the middle of the front yard and stood it where I would have a clear view of the variable stars in the southeastern sky. That was the night that I forgot all about telescopes and variables for as I turned and looked up at the sky, right there in front of me—squarely in the center of the Milky Way—was a bright and blazing star!

I have always wished that I could recapture my sensations of those first few minutes of that sighting. That I was bewildered and confused goes without saying for I had acquired a fair knowledge of the stars and constellations and here, right before me, was a total stranger, a star that had not been there just the night before. I do remember wondering, momentarily, if one of the planets could possibly have strayed that far from the ecliptic. I checked the star's position in my Upton's *Atlas* and my *Field Book of the Stars* still has my penciled plottings of that night.

It was my first view of a nova, and what a wonder of a new star it turned out to be. Nova Aquila—as it was called—was, when I first saw it, of equal brightness to its near neighbor Altair, which is a standard first-magnitude star. Before the night was over it was even noticeably brighter than Vega, the brightest star of the summer sky. It was seen that night all over the world by people in all walks of life who were familiar with the stars. It is logical to assume that it may first have been seen by shepherds or by wandering nomads in the Far East, where darkness came earlier than here. Several of the astronomers still at their eclipse sites saw the star when darkness came to America. Barnard, of Yerkes, who during his lifetime seldom missed any newcomers in the sky, saw it from his station at Green River, Wyoming, while on his way back to town after packing up his plates of the eclipse.

Following the announcement of the appearance of Nova Aquila, quite a number of people insisted that they saw it bright on the previous night. Fortunately, the eye of the patrol camera had also seen the nova on that night and had recorded it as being of the sixth magnitude—just barely visible to the naked eye. Thus, even in 1918, such things as vivid imaginations, poor memories, and notoriety seekers already were abroad in the land.

All during the night of June 8, the nova increased in brightness and the following night reached its peak. It was by far the brightest nova in more than three hundred years since 1604 when Johannes Kepler, the great astronomer-mathematician, had watched a similar outburst in the nearby constellation of Ophiuchus. My own estimate of the nova on June 9, when at its greatest brilliance, was magnitude, minus 1.5, or almost exactly the equal of Sirius, the brightest of all the stars. Since Sirius is a winter star and not visible in June, I could only make this estimate by my memory of Sirius as it appeared in the winter sky.

This is decidedly NOT the way to make estimates and I was greatly relieved when the final official figure of its brightness was announced as magnitude, minus 1.4 or one-tenth magnitude fainter than my guesstimate.

Following its brief night of glory the nova began to fade, quite sharply at first, then more gradually and with many minor fluctuations until, about eleven years later, it had finally simmered down to its original twelfth magnitude and there it has quietly slumbered ever since, content with its dreams of that night in June when it came on stage—a newborn star—and stole the show!

12 Copus Hill

ON NOVEMBER 11, 1918 WORLD WAR I FINALLY CAME TO AN END. Whistles blew, bells rang, there was great celebrating, and Alfred Noyes penned the satirical lines of "A Victory Dance." After passing the winter with the Army of Occupation in Germany, Kenneth returned home, was married shortly after, and then moved onto his own sixty-acre farm adjoining ours on the north. My sister, Dorothy, also married, was now living in a nearby town.

Suddenly our six-room house had become quite commodious and I now had a room all my own. Like a child with a new toy I made the most of it. I painted the walls and ceiling and hung new curtains at the window. I built a long, narrow workbench which I placed in front of the window and in the center of this bench, where it would get all the daylight possible, reposed my homemade microscope. This was a miniature of a regular model and I had fashioned it of walnut, using for its optics, the 60-power eyepiece from my spyglass. At one end of the workbench, out of the direct light, stood a small aquarium filled with pond water from the woods. It was what my biology textbook called a balanced aquarium for it contained a self-maintaining balance of both plant and animal life. Duckweed, like little lily pads, floated on the surface, while below, tiny daphnia and bristle-tailed cyclops spun around in jerky orbits. Half a dozen pale, translucent fairy shrimps, with undulating plumes, swam about serenely on their backs while two scuba-diving beetles, their spheres of air gleaming like drops of mercury, searched for sunken treasure among the plants and seashells on the sandy bottom. As custodians, four bright-red snails, in true slow motion, plied their sanitary push brooms in crisscross streaks about their little home.

On the east wall of my room I fastened the walnut case that I had made the year before to house my collection of butterflies and moths. This case, when first made, had a tightly fitting door but later a tiny crack developed somewhere and through it my case came to house just one insect too many—a little matron named Anthrenus—who laid her eggs in the bodies of my other specimens and before I knew it I had nothing but a pile of assorted legs and wings in the bottom of the case. I puttied up the crack in the door and after disinfecting the case with carbon disulphide I repaired my old insect net and started making a new collection. Cocoons collected on my winter walks, larvae raised in wire cages in an unused shed, and a couple of nights spent in "sugaring" a tree back in the woods by painting the trunk with an appealing concoction of honey and syrup soon repopulated my case with a new world of Lepidoptera.

The north wall of my room held two smaller cases. These had no doors but consisted simply of closely spaced open shelves. In one case was my collection of local rocks and minerals together with a number of fossil brachiopods that had been washed out of the gravel of the river bank. The remaining case held Indian relics—arrowheads, stone axes, and a couple of skinning implements of flint—nearly all of which had been picked up on our own farm. In one corner of the room a hornet's nest, roughly a foot in diameter, hung suspended from the ceiling, while on a shelf above my bed were about a dozen of the books that currently rated highest on my most-often-wanted list.

Books were never easily come by except, of course, at Christmas and on birthdays but with each book that I bought I followed my mother's commendable practice of inscribing not only my name but also the date of purchase on the flyleaf of every newly acquired volume. As previously intimated I save things, and I still have all those books that I bought in my declining teens. Most of these, of course, deal with astronomy—Todd's *Stars and Telescopes,* Flammarion's *Astronomy for Amateurs,* Olcott's *Star Lore of All Ages,* Serviss' *Curiosities of the Skies,* Webb's *Celestial Objects for Common Telescopes,* McKready's *Beginner's Star Book,* and *Pleasures of the Telescope* by Serviss all have teen-age dates in them.

On other varied subjects for these same years were such titles as *The Book of Woodcraft* by Seton, *The Butterfly and Moth Book* by Robertson-Miller, *Twenty-Thousand Leagues under the Sea* by Verne, *Minerals and How to Study Them* by Dana, and Darwin's *The Voyage of the Beagle.* These early dated books were purchased with my own infrequent dollars and thus they tell me, as well as could the pages of a diary, what were my chief concerns in those first two postschool years—though I will cheerfully admit to still other interests and other dates than those recorded by my books.

At the time I left school it was planned that with the coming of peace and with my brother's return from service I would go back and finish my final year of high school. These two conditions now were both fulfilled and, in addition, we now had a car—a Model

T—which would have solved my old problem of transportation. But now I had not the slightest desire to go back to school. In fact, when I left school I was fairly certain in my own mind that I would never return. Nevertheless, during the summer I gave this a lot of thought and finally decided that a high school diploma would be of little benefit unless I went on to another four years of college. This was not even to be considered. Not only would it have been a financial hardship but there was still just as much work on the farm as ever.

To be completely honest, there was yet another reason why I took a rather dim view of any further schooling. That reason was that life on the farm was so pleasant, so independent, and so complete that I had no desire to give it up. By this I do not mean that it was all fun, for certainly there was plenty of hard work and long hours, especially during the rush seasons of planting and harvesting but it was labor and time devoted to one's own interests. My parents left the matter entirely up to me and never once have I regretted my decision.

Unless one of our rush seasons was on I usually managed to get back to the woods nearly every day—most often in the early evening. On Sundays I would spend the entire afternoon either in the woods or along the river. In summer I would be armed with my butterfly net, a cyanide jar for killing specimens, and a small stiff box for storing them until I reached home. In winter, with plenty of leisure time, these walks were often all day treks and I then ranged far afield. On such occasions I always took along a lunch consisting of some bread and a large slice of sugar-cured ham which I would broil on a green stick of wood over a small fire.

In my nineteenth summer I took my first vacation camping trip. Dad and Mother had just returned from visiting some relatives living near Mansfield, Ohio, about a hundred miles east of our home. While there they had hunted up the location of what was known as "The Copus Monument"—a small marble shaft that had been erected to mark the site of a battle in 1812 in which Mother's great-grandfather, James Copus, and three soldiers were killed by a band of marauding Indians.

Their little pilgrimage to those hills so steeped in the early history of our family had so impressed Mother and Dad that on their return home they urged me to take the car and make the same trip. It was a most welcome suggestion and I began to make plans at once. Today it seems quite incredible but, at nineteen, I had never been away from home overnight and at that time my furthest peregrination had been made as a child of six or seven, when my mother and I had made a one-day trip on the electric interurban car to Fort Wayne—fifty miles away!

As my companion on this expedition into the wilds of north-central Ohio I decided to ask my good friend Gilbert Miller. We had become acquainted in high school and the friendship had prospered to the point where we got together at least once a week. This was usually on Saturday night, when we would either play records on his victrola or go to an early movie. Gilbert was an extremely well-read chap whose mind was a vast and orderly storehouse of facts about a great diversity of subjects—particularly on things scientific and mechanical. In addition, he was a born musician, with an ear aware of every nuance of tone and with the facile touch to faithfully interpret it. I have greatly benefited by our long acquaintance. Through him I first met Omar Khayyam; he introduced me to the hilarious humor of P. G. Wodehouse; and from his hands came my first hearing of the rich tone poems of MacDowell's "Woodland Sketches" and the "Peer Gynt Suites" by Grieg.

Gilbert was just as delighted as I at the prospect of a camping trip and we decided on a stay of at least four days. We would travel, of course, in our "tin lizzie," as these early Fords were more-or-less affectionately called. This car had made possible our weekly get-togethers, for it was mine for the asking nearly any night but lodge night, and Gilbert and I both were familiar with its virtues and its limitations. In those days a new car came equipped with a full set of tools, including jack and tire pump, and one was supposed to know how to use them. There was no spare tire, for neither the wheels nor the rims could be taken off the car so all tire repairs had to be made on the road right at the scene of the disaster. Of necessity we had become fairly familiar with the intricate innards of our Model T. We had ground valves, installed piston rings and relined transmission bands as well as performed numerous minor operations. Our car was a five-passenger touring car—one had a choice of that or a two-passenger roadster—and as for color there was even less variety for, as Henry Ford once said: "We'll give them any color so long as it's black." But we were proud of it. It was a good car and worth every cent of the three-hundred and fifty dollars we had paid for it.

Early on an August morning we started out. The rear seat of the car was piled high with blankets, pillows, and extra clothing. One noisy carton held our dishes, our frying pan, and our coffee pot, while a large covered hamper was filled with our provisions. We left well stocked with all those edibles that a camper or a pioneer traditionally consumes. We had enough bacon and eggs for a two-week tour and our supply of coffee would have warmed the heart of any Brazilian Chamber of Commerce. We had no time-honored haunch of venison but we did have a great quantity of thick slices from a sugar-cured haunch of hog. Included also were numerous cans of pork and beans and Gilbert's mother sent along a big batch of homemade cookies and doughnuts. As I recall it we had no salads or green vegetables of any kind—but probably neither had our predecessors, Daniel Boone, and Lewis and Clark. As we left Dad remarked that maybe we ought to take along some beads and cheap trinkets as trading stock just in case we met up with any Indians still lurking in the wilderness.

During most of the long drive we bowled right along at twenty-five to thirty miles an hour for the roads, though still unpaved, were dry and in fairly good condition. We soon left the flat farmland behind and the roads led up and down over rolling hills devoted to grazing and often covered by thick woods of oak and beech with here and there an occasional yellow birch to tell us of the lessening lime content of the soil.

As we approached Mansfield the terrain became even more rugged. These were the first real hills that we had ever seen and they also seemed to be the first that our Model T had yet encountered. It, quite frankly, didn't care for them and several times it protested with considerable warmth. Going down hill, though, was a lot of fun. Once headed downward on a long hill I would turn off the switch on the coil box—the car had neither battery nor instrument panel—and we would coast all the way to the bottom. Here I would turn the ignition back on, throw in the high-speed lever, release the clutch pedal and the engine would start again. Unfortunately we could never use the momentum acquired in coming down to help us to climb the succeeding hill for there was always a rickety one-way bridge at the bottom that slowed us down to a crawl. If the ascent was a steep one, long before we reached the summit I would have to push the clutch pedal all the way down. This threw the planetary transmission—the pride and joy of the Model T—into low gear with the result that the radiator cap soon became a miniature of Old Faithful as clouds of steam were wafted back to us through the opened upper half of the windshield.

In the little town of Mifflin we drove slowly, watching for a pump and horse trough on our right and here, following the map that Dad had drawn for us, we turned off the highway onto a narrow country road. After several miles of cautious winding in and out among the wooded hills and after fording two small streams we finally sighted—directly before us—a sharp white spire rising above the surrounding low bushes. It was the Mecca of our journey—the Copus Monument.

We stopped the car on the hillside road just above the monument and looked about. Along this road, on the afternoon of September 15, 1882, just seventy years to the day after the battle, more than 1,200 horse-drawn carriages, wagons, and carts had passed in review at the unveiling of the monument. We read the inscriptions on the shaft then drove slowly on. The afternoon was nearly gone and we still had to find a suitable campsite before dark. Less than a mile further on we found it—a perfect little spot right beside the rocky bed of a rushing little stream. At the farmhouse just around the next bend in the road we got permission to establish our headquarters and going back to the stream we let down the bars of a pasture gate and drove right up to a patch of sand alongside the water.

Setting up camp was a simple chore. While daylight still lingered we collected a great heap of firewood from fallen branches in the woods nearby and from piles of driftwood that had lodged along the stream. For our first supper we opened a can of pork and beans and heated them right in the can. We ate the remaining pieces of cold fried chicken and sandwiches that my mother had sent along for our lunch and then finished up with doughnuts and a pot of coffee. Throughout our stay our only source of water was the crystal stream beside our camp. We boiled the water thoroughly and it was safer than that from any spring or farmhouse well.

We lingered a long time over that supper. We were tired and hungry after the long excitement of the trip and food had never tasted better. Even that lowly can of pork and beans seemed somehow to be seasoned with the aura of our strange surroundings. As we ate we watched the nearly full moon clear the treetops on the high hill to the east. In a nearby thicket a mockingbird rehearsed its entire repertoire then finally an owl, somewhere in the gloom, began its round of questioning.

We washed our tin dishes in the stream and dried them by passing them through the flame of our campfire, thereby sterilizing them as well. We got out our blankets and made up our beds on the sand right beside the car, but before I turned in I got my spyglass out of the car where I had it wrapped in an extra pillow. For weighty reasons I had left the grindstone mount at home but the car top made a fairly steady support for the instrument. The southern sector of the sky being clear of hills and trees I made a long and careful search just above the horizon between Scorpio and Sagittarius on the chance of picking up Finlay's periodic comet which had been lost since 1905 but now was due again. In the bright moonlight it was, at best, a forlorn hope and I saw only vaguely the great star clusters of that region so I decided to wait until I had a steady mount and a moonless night. But I was able to make estimates of Nova Aquila and three irregular variables, R

Coronae, R Scuti and RY Sagittarii. I found them all between sixth and seventh magnitude and thus easily seen in spite of the brightness of the night. The stars all seemed to be on their good behavior so I put away the telescope and went to bed.

Already Gilbert seemed to be sound asleep but I followed the suggestion he had made a short time before. I crawled between my blankets, lay on my back and then wiggled about with a hula-like motion until I had made a perfectly contoured mattress of the dry sand underneath. What an ideal place, I thought, for watching meteors and for a time I lay awake hoping to see a few. But the Perseid shower had now been gone for an entire week and in the bright moonlight I never saw a single one. My final contribution to science for that day was the drowsy observation that Vega was so nearly overhead that if she were to fall she would surely land here in our little valley. And the last sounds to come to me on that day of high adventure were the babblings of our bedside brook and the distant baying of a coon dog.

Toward morning I awoke. I was cold, and reaching out, replaced my blanket. The perfect contour of my evening couch had also disappeared for I now was lying on my side. With another wiggle that too was easily restored. While I slept the moon had crossed the sky and now was slipping down behind the distant, wooded western slope. Cassiopeia had chased Vega far to the west and now was nearly overhead. Approaching dawn outlined the trees of the high eastern hill and standing in the topmost branches was Orion. Scattered out across the east were the Pleiades, Aldebaran, Capella, The Twins, and Procyon. Just out of sight, hidden by the hill, was Sirius. Again, I had a preview of the frosty winter stars.

Right after an early breakfast of bacon, eggs, doughnuts, and coffee we drove back to the monument. White wisps of fog still lay in the lowlands on either side of the narrow rocky road and covered the broad valley with a downy blanket. On our right the high wooded hill, backlighted by the rising sun, was emerging from the mists of dawn. We stopped the car on the sloping hillside trail—where Gilbert, as a safety measure, blocked one rear wheel with a stone—and walked the few remaining yards down to where an ornamental iron fence did picket duty for a slender marble shaft resting on a heavy granite base.

On the upper portion of the base was engraved the name James Copus. This was followed by the names of the three soldiers killed in the same battle. Also, as a memorial, the name Jonathan Chapman had been engraved. Chapman, better known as Johnny Appleseed, died and was buried near Fort Wayne, Indiana, in 1845 but he frequently had visited the Copus cabin and was a trusted friend of both Indians and whites.

On our August morning the scene that lay before us must have appeared much the same as on that fateful morning in September, 1812. The same tree-clad hills, even some of those same trees, looked down on a little clearing that once had held a small log cabin and a barn. The cornfield that we saw in the background might well have been the field of ripening corn in which forty-five Delawares lay hidden that early morning until the soldiers left their quarters in the barn and strolled, unarmed, over to a nearby spring. As related in Henry Howe's *Historical Collections of Ohio,* at the first war cry James Copus seized his musket and stepped outside his cabin door only to exchange fire with a savage standing a few yards away. Both fell mortally wounded. In the battle that ensued three soldiers also were killed and only the timely arrival of reinforcements from a nearby blockhouse averted the almost certain massacre of Mrs. Copus and her seven children and the six remaining soldiers.

Gilbert and I spent the next two days exploring the hills and valleys adjacent to our camp. We followed our rocky stream far along its crooked course and delighted in its quiet pools and tiny waterfalls. The glacial terminal moraine winds irregularly through this region and everywhere about us we could see erratic boulders that had been dropped from the melting frontal edge of that great ice sheet which covered much of Ohio some fifty thousand years ago.

On one of these hikes we came upon three young killdeers who apparently had recently severed home ties and now had started out to see the world. They were about half-grown and could run quite well but still were incapable of any sustained flight. Gilbert managed to capture one of them and after a little gentle petting it seemed to lose all fear of us. I had my camera along—a small Brownie box camera equipped with a portrait lens attachment that gave a larger image by taking everything from a distance of three and a half feet. Gilbert placed our young feathered friend on top of a large granite boulder, turned him broadside to the sun, and I clicked the shutter. The eventual result was purely a case of beginner's luck for when the roll of film was later developed and printed our little bird stood out clear, sharp, and lifelike. I sent the print to the editor of the Nature and Wildlife Department of American Photography Magazine and a few months later I was overjoyed to find the beady eye of our young killdeer regarding me from the printed page. I thus was fairly launched on another lifelong hobby.

On our final night in camp we were awakened about one o'clock by big raindrops spattering in our faces. Hastily we rolled up our bedding and tossed it in the car. Then we dug out the two left-hand side curtains and fastened them on the windward side of the car. We had barely finished this before the warning sprinkle became a downpour. Crouched in the front seat, wrapped in blankets, and buffered about with pillows we soon drifted back to sleep, lulled by the rolling thunder that bounced about among the hills.

When we awoke in the morning our friendly stream had become a torrent, our bed of sand was under water, and our store of firewood had drifted away. So we packed up and started for home, taking a longer route back to the highway to avoid the streams now too deep to ford. We soon stopped at a restaurant for breakfast and, for some reason, neither of us ordered coffee.

Our return trip was uneventful and our arrival an anticlimax. It seemed that I had been gone for weeks and many new and exciting experiences had come my way. But at home nothing had changed. The house still needed painting, the same five cows had to be milked twice a day, and, as Mother quoted, we still had the same old cat.

It had been a wonderful trip and Gilbert and I solemnly vowed that we would return to Copus Hill year after year. But somehow we never have. Too many other things have needed doing. In our reminiscent moments we often have recalled the wonders of that trip and wished that we could relive it all again. Only last year I suggested to him that we make another try. This time he did not eagerly accept as on that first occasion. In fact, he did not accept at all—he refused.

"Les," he sagely observed, "you forget, we aren't nineteen any more. We can never really go back to something like that was. There have been too many changes. We would only be disappointed in it now."

He was right, of course. I still was looking at a teen-age mental image. I had only to observe the changes all around me to conjure up a realistic picture of what those hills must really be like today. We would doubtless find those big oak and beech trees timbered off and the gentler hillsides cleared for contour farming. New roads would have been built and the creeks that we forded cautiously have by now been bridged with concrete culverts. Even the stream beside which we camped has doubtless now been dredged and we wouldn't drink its water now if it were boiled all day. Suburbia has, by now, moved in and a ranch-type house must squat on every level bit of ground. The hills are laced together with telephone and power lines and their once-quiet slopes now echo with the whine and sputter of the power mower. Only the stars I saw that night are still untouched by man.

13 Cow Pasture Station

THE SLOWLY DECLINING NOVA AND MY CONSTANTLY GROWING observing list of variable stars kept the 2-inch busily occupied for many months. During the fiscal year of the AAVSO which ended in September 1919 it had watched the stars on a total of 190 nights—more than half the nights of the year. As a result of this the spyglass worked itself right out of a job. In November the Telescope Loan Department of the AAVSO offered me the use of a 4-inch telescope.

This department had recently been established as a means of providing the interested observer of variable stars with a more adequate telescope than the one then in use. Some of the instruments in their collection had been gifts to the Association, others were bequests, a few were outright purchases, while still others were on loan to the AAVSO from former observers who were now inactive. The only stipulation attached to the loan of one of these instruments was that it be kept in reasonably regular use in the observing of variable stars. Quite naturally I was delighted with the offer they had made and accepted at once. The 4-inch arrived early in December and the strawberry spyglass was retired to a place of honor on a shelf in my room.

The new telescope was a Mogey refractor of about 60-inch focal length. It had no tripod but was equipped with an equatorial bearing designed to be placed on top of a permanent pier or post set in the ground. This meant the end of that era of moving about from place to place in order to avoid trees and buildings. On the other side of the ledger, of course, this also meant that on windy nights I no longer could observe from the sheltered lee of the house. And I could also foresee a gradual softening of those muscles which, for three years, had lugged the grindstone mount to various strategic spots about the yard.

I was not long in deciding just where to locate the permanent observing station that would now be necessary. We had the perfect spot for it, right in the center of a small pasture field directly west of our house. At this point there was a good clear horizon all around except low in the east where the elms and maples around the house rose to a height of about 20 degrees. From this vantage point there would be no lights to interfere from any direction, not even from the occasional automobiles that passed along the country roads. Here I would be only about a hundred yards from the house and this added advantage of easy accessibility, I knew, would be particularly appreciated in winter whenever I had to shovel a path through a heavy fall of snow.

The telescope had arrived during an extended spell of gloomy weather and for once the clouds were welcome. They would give me an opportunity to get the scope properly mounted before a starry night came along. First of all I had to have a pier on which to mount the equatorial bearing. I walked back to the woods and soon located a straight, young, white ash tree about seven inches in diameter. I cut this down and then chopped off an eight foot length and carried it back to the barn. Here I peeled off all the bark, sawed both ends off square, and then bolted the bearing to the top end of the post. Carrying the post out to the pasture I placed it on the ground while I carefully selected the precise spot on smooth, level ground, to erect my pier. It had to be fairly near the house for convenience, but still not too close or the trees around the house would interfere with the seeing toward the east.

While I was engaged in all this reconnoitering a couple of our cows, seeing me, left off their nibbling at the strawstack near the fence and ambled over where they took up strategic positions and watched my every move. On the selected spot, after chopping through a crust of frozen ground, I augured out a hole three feet deep, lowered into it my white-ash pier and wedged it solidly in the oversize hole with three strips of wood. "Now," I told myself, "I'm ready for the first clear night."

I had not long to wait for on the following evening the sun set in a clear blue sky. Darkness falls early in mid-December and by six o'clock I had the telescope bolted to the mounting. I stood on my three-step stairway and swung the scope toward Vega, hanging in the low northwest. Three years before, on a warm summer evening, she had been the first star for my 2-inch spyglass. In the 4-inch she was almost dazzling and, after critical focusing, beautifully sharp and clear. Being so close by I next looked at the Ring Nebula and, though it was low in the sky, I clearly glimpsed the dark hole in this celestial "doughnut." Just above Vega was Cygnus and in this constellation I paused for a glance at one of my favorite irregular variables—SS Cygni—and was delighted to find that I could easily see it at its minimum brightness of 11.9 magnitude. Fired with enthusiasm I swept south along the Milky Way and soon located the field of another old friend, the variable star with the ten-thousand-fold range in brightness—Chi Cygni. I could just catch glimpses of it near its minimum at magnitude 13.

For the next hour I wandered here and there about the sky getting acquainted with my new scope and with the completely new and glorified appearance of many of the old favorites that I had watched so often with the spyglass—the Pleiades, M35 in Gemini, M31 in Andromeda, the Orion Nebula and its quadruple star Theta, then the blood-red variable R Leporis and the orange and blue colors in the double star Albireo.

The night thus far had been devoted entirely to sight-seeing. As yet the pier of my mounting had been only temporarily positioned and the equatorial bearing only vaguely pointed north. This being my first astronomical telescope, my first equatorial mounting, and my first permanent station I went to great length to get everything as accurately adjusted as I knew how to make it. First I loosened slightly the three wooden wedges that held the pier tightly in the hole. Then, by twisting and tilting the pier, I maneuvered it so that the polar axis of the mounting seemed to my eye to point directly to the celestial pole—a vacant spot about a degree from Polaris on a line leading toward the end of the handle of the Big Dipper.

The great advantage that an equatorial mounting has, over one such as my grindstone affair had been, is that with the equatorial the polar axis is parallel to the polar axis of the earth and thus only one motion—either a downward pull or an upward push—is required to keep a celestial object in the field of view. If the equatorial has been properly adjusted one can follow a star all the way across the sky from east to west using this one motion only and this, in principle, was the way that I tested the setting of my mount.

I selected three stars having the same declination, or distance from the celestial equator. For the greatest accuracy these stars should be widely spaced—one in the east, one south, and the third in the west. Knowing that the star Delta, in Orion, was almost exactly on the equator it became my eastern star. My atlas showed that a faint naked-eye star, 60 Ceti, was also on the equator and directly south. The third equator star, and located in the west, was Zeta Aquarii. Next, in my big copy of *Revised Harvard Photometry* I looked up the accurate positions of these three stars and found them to be spaced 3½ hours apart along the equator while in declination they were respectively 22, 21, and 31 minutes south of the celestial equator. This made a difference of less than one-third the apparent diameter of the moon. This meant that when my mounting was properly set I could get Delta in the center of my field and clamp the scope in declination, then move on westward to 60 Ceti and watch it cross the field—also virtually in the center. Lastly, I would move on to Zeta in the west and find it one-third of a moon higher in the inverted field than the other two had been.

Needless to say it took quite a bit of twisting and tilting of my pier before I achieved this desired accuracy but when I finally had it all lined up I left it there that night and came back the following morning with a bucket filled with a soupy mixture of sand, water, and cement which I poured into the hole around the pier. All this toilsome striving was, of course, quite unnecessary for anything other than long-exposure photography. I had, without a doubt, the most accurately adjusted white-ash equatorial mounting in the annals of astronomy. But for variable stars, where each observation requires only a matter of seconds, an old discarded grindstone works just about as well.

In actual practice, when each night's observing was finished the telescope tube was removed from the mounting by unscrewing two thumb nuts and the tube was carried into the house. The mounting remained on the post and was covered with a heavy grain sack to protect it from snow and rain. Close by the pier I set another, but much shorter, post in the ground and on top of this I mounted an accessory invention—an open-air desk. This strange but practical device was a large weather-tight box about three feet long, two feet wide, and one foot deep. The lid of the box was a framed pane of glass. Inside I kept my atlas, charts, and red bull's-eye oil lantern. Here everything was easily visible through the glass top and also readily accessible so that I could turn pages and record observations through the hinged front panel of the box. The whole affair was pivoted on top of the post so that it could be turned in any direction to avoid the wind.

When spring came I found it necessary to build a fence around my open-air observatory for it was located in our cow pasture and the several members of our local galaxy, exercising their prior rights, continued to make free use of the observatory grounds. Before I could get the area properly enclosed they had even started to use my meticulously positioned pier as a rubbing post. From many nocturnal hours spent in their company I learned that cows are light sleepers. Several times during the course of a night they would rise, eat grass for a while, then lie down and ruminate on the results of their grazing, then take another nap. Cows are friendly folk and have a great sense of curiosity. They often would come over and keep me company during my long hours at the telescope, standing just outside the fence, their heads hanging over the low top wire, slowly chewing their cuds and watching me thoughtfully with their big soft eyes.

Cows were always my favorite farm animals and until my early thirties there was always a cow in my life—usually several of them. In our very early teens my city cousin and I often would ride our more sedate and docile cows here and there about the barnyard. We had no guidance system of any kind and our progress was subject entirely to the prevailing mood of our mount and any spirited urging on our part always terminated the ride quite suddenly. I may at times long for the good old days back on the farm but I have retained no nostalgic yearning for a repetition of any of these bucolic capers. The cow is ill designed for bare-back riding. Their backs are much too sharp for comfort and, having no mane, they offer absolutely nothing to hold on to when their patience with young cowpokes finally becomes exhausted.

We always had two or three cats on the farm and at chore time if they saw me start for the barn carrying a couple of milk pails they would fall in line and trot along behind, their tails waving aloft in anticipation. Arriving at the scene of operations they would line up expectantly in a row nearby. At the first sound of the milk striking the bottom of the pail they would rise on their haunches and, with forefeet pawing the air, they would begin to mew. I would reward each one with several squirts of milk directly into its wide open mouth. Then, when they realized that the show was over, they would sit and lick each other's faces and wait until I was finished, when I would give them some more milk in their pan.

My dairy days, of course, came long before the advent of the antiseptic cow, and any modern dairy farmer with his shiny milking machines, his surgically clean stalls, and his siring subterfuges would stand aghast at some of our primitive practices. But cows were cows in those days and, somehow, I feel sure that life had a deeper meaning for those rugged matrons of yesteryear than it has for the immaculate robots of today.

One of our cows, a temperamental Jersey, was a confirmed kicker. The only way she could be safely milked was to fold up her left hind leg at the middle joint—anatomically this was her heel or hock though we always called it her knee—then we would slip an iron ring over the folded joint. To perform this very simple operation you grasped the cow's nearest hind leg with your left hand, at the same time throwing her weight on the opposite leg by pushing hard with your head against her flank. While she was still off balance it was easy to pull up her left leg and slip the ring into position. This would leave her left leg dangling in mid-air and the cow in a state of unstable equilibrium. Obviously, three-legged cows do not defy the law of gravitation by kicking.

Throughout the summer season our cows would be out on pasture every day and with the approach of fall they would come in for the evening milking with their tails a solid mass of burrs. To milk a cow equipped with one of these free-swinging fly swatters was truly an occupational hazard. The cautious milker either tied the tail to the cow's leg or held it tightly clamped in the bend of his own knee, for if this burr-bedecked bludgeon ever landed on his cheek or the back of his neck some of the barbs invariably remained there, firmly embedded, as a lasting reminder of the efficacy of one of nature's marvelous methods of seed dispersal.

Throughout the ages the cow has been mankind's most useful animal yet, at the moment, I can recall no individual bossy who has ever left her cloven hoof print in the long muddy pavement of history. Even that remarkable cow who leaped through outer space into the first lunar orbit was but a nameless cow.

In the varied menagerie of the sky my faithful friend the cow has been accorded the same shabby treatment as here on earth. Had I been in charge of designing and naming the constellations I would certainly have bestowed a place of honor in the zodiac on the cow instead of on her huffy husband Taurus, the Bull. The celestial vault is bestrewn with bears, horses, birds, and lions. We have four dogs and three fishes. We have a dragon, a scorpion, and a lizard. There is even a fly in the firmament. But the patient cow does not even get credit for the Milky Way!

14 We Build an Observatory

FROST AND HEAVY DEWS WERE THE MOST SERIOUS HARASSMENTS that I encountered in working in my open-sky observatory. In spite of a protective sleeve extending over the lens, on some nights the cold surface of the objective would cloud up perhaps half a dozen times during an evening's work. Dew, of course, could be removed with soft, absorbent tissue, but when frost was the offender I would carry the telescope back into the house, let it warm up thoroughly, then take it back out and work until the dimming star images would tell me that it was time to defrost again.

I am quite certain now that there were better ways of doing this. I am sure that I could have devised a somewhat elaborate extension tube over my lens—perhaps with an inner coating or filling of calcium chloride—which would have kept my objective dew- and frost-free for many hours. But I never did and the only excuse I can offer for persisting in this time-consuming practice of warming the telescope by the fire is that it provided an often welcome opportunity for the telescopist to warm up a bit at the same time. Standing in an open field for a couple of hours on a midwinter night can be a bone-chilling experience. Sometimes on these frigid nights I would briefly leave the telescope and run around and around my little fenced enclosure until I had partially restored my lagging circulation and brought a feeble glow of life back to my icy hands and feet. During the months of March and April the occasional windy nights were also somewhat troublesome for the tremor of each little gust was magnified by the power of my scope, but for the balance of the year any daytime breeze we had usually died down with the approach of darkness. It was rare indeed that I encountered any noticeable wind in the hours after midnight.

I used my observing station on nearly every clear night for two wonderful years and, in spite of any exasperations of frost, dew, and wind, I still feel that the open sky is the best of all observatories for the beginning amateur. Under the open dome of sky one is at least half surrounded by the stars. Stars are above us, stars sparkle on all sides of us down to the horizon. We are actually between any two diametrically opposite stars and this seems to make us a vital part of the hemisphere of visible sky and not merely a spectator looking out through a narrow window.

I am truly thankful that I did not miss these years of observing in the open, for the impressions of that period that I most vividly remember today are, for the most part, those little perceptions which I never would have known had I been surrounded by four walls. Out in that open field all the sounds of night seemed close about me as the seasons slowly clicked their way through the turnstile of the year. The sharp, dry creak of snow, the vernal "sweet ode of the tree toad," the night heron's strident quonk, the shrill alarm of a nesting killdeer disturbed by some night prowler, the rhythmic beat of the snowy tree-cricket's thermostat, the melancholy hooting of a great barred owl, and, finally, their haunting clamor dropping from the darkness overhead, the southward winging skeins of geese.

This open-sky period occurred at a time of considerable sunspot activity for I remember so well some of those brilliant displays of northern lights. At least two of these I have never seen equaled since that time. Unforgettable sights they were, with great shaking curtains of multicolored light so bright they cast distinct south-pointing shadows on the snow. As I watched I always listened for the crackling sounds which, it seemed, should be a part of such celestial static. I always hoped for something audible to accompany that eerie splendor but it never came.

In such work as variable-star observing a dark sky background is absolutely necessary in order to distinguish faint stars, so one avoids the moon as much as possible. Whenever the moon was fairly bright in the evening sky I would go to bed earlier than usual after setting my alarm clock for the time of moonset as listed in our *Farmer's Almanac*. Then, when it awoke me, I would get up, dress according to season and temperature and make my way downstairs in the dark. Here I would pick up the telescope tube from its place in the corner of the dining room and carry it out along the path—sometimes a dim trail worn in the grass, sometimes a dark line cut through the snow—until I came to the middle of the pasture where a tall post loomed against the sky.

In following this weird nocturnal schedule required of a moon-dodger I found that all the sounds of nature are at their lowest ebb in the cool hours shortly after midnight. The tree crickets never seem to quite give up during summer nights although the cold of dawn greatly slows their tempo. Of the birds—so voluble in early evening and at sunrise—only the catbird and the owls have I heard late at night. The tree-toad chorus also gradually dies down though sometimes a distant bullfrog has kept me company throughout the night. It always amazed me how early the approach of a summer daybreak could be detected but even so, here in the country, there always was a rooster somewhere ready and waiting for it and his first clarion call would soon be answered from other farms in all directions. Tennyson's line: "The earliest pipe of half awakened birds," must certainly have referred to the domestic variety.

Early in the fall of 1921 Dad decided that I needed an observatory—a regular observatory, dome and all. Just what brought about this sudden decision I have no idea. It was certainly no hint or suggestion of mine for I had been perfectly happy with my open-air station. Nevertheless it all sounded fine to me for I felt that it would relieve every inconvenience of the present setup. We started right in, though neither of us had ever seen an observatory. We decided to make the building 10 feet by 14 feet in size, with a 9-foot diameter dome. This would give plenty of working space beneath the dome and provide an alcove 4 by 10 feet in size for a desk and other equipment.

We wasted no time by drawing up any plans. Dad was a good practical carpenter and had built our six-room house without any set of plans so I had no worries as to the outcome of the present project. The location for the new observatory was staked out right beside the open-air station in the cow pasture for this had proven, after two years' use, to be nearly ideal. A concrete mixer was borrowed from a neighbor and, with Kenneth's help, the foundation and the footing for the telescope pier were installed in one day's time. I must have had some delusions of grandeur regarding that pier for we put in a solid cubic yard of concrete level with the surface of the ground and in this we placed an upright 6-inch steel pipe to hold my 4-inch refractor. I had never heard of any Ohio earthquakes but I was prepared for them.

The walls and the roof went up in short order but the dome presented more of a problem. The circular track on which the dome was to revolve, as well as the base of the dome and its curved ribs all were built up of laminated three-quarter-inch boards all sawed out to the proper radius—and we had no bandsaw! It all added up to no less than seven hundred lineal feet of sawing, all done by hand with a small compass saw. Fortunately we used cottonwood and pine for this rather than oak.

After the framework of the dome was completely assembled it was covered with light-gauge galvanized iron. Then Gilbert and I, with considerably more enthusiasm than skill, spent a couple of days in soldering all the seams together to make a solid weather-tight unit. The completed dome was then hauled out to the waiting building and carefully set in its appointed place. Everything fitted perfectly and a gentle push on its short handle just above my head easily revolved the dome in either direction.

The observatory housewarming was a solemn occasion. It now was January and the special guest of honor could not arrive until three o'clock in the morning. I set the alarm clock with care and promptly on the hour the dome was turned to the northeast, Vega made her sparkling entry through the telescope and all was well.

I built a little table for my books and charts and, as electricity still had not come to the farm, I rigged up a tiny ruby lamp that operated from a dry cell and this served to illuminate my atlas and my star charts. On the walls I hung two pictures. One was a small portrait of Edward C. Pickering, then director of Harvard Observatory, who each month signed a note of thanks for the variable-star observations that we sent in. The other picture showed old Galileo clutching in his hands the little spyglass that had first revealed the satellites of Jupiter and the craters of the moon.

The dome proved to have just one minor flaw in its otherwise efficient operation. The sheet-metal shutter that covered the opening in the dome had to be removed by climbing up on the roof by means of a ladder outside the building, then lifting off the shutter and carrying it down to the ground. Then, of course, when the observing session was over it had to be carried up the ladder again and put into place. The shutter was not heavy and I soon became accustomed to this ritual and didn't mind it at all. In fact, I was quite pleased with my acquired ability to hold the shutter above my head with both hands and walk down the inclined ladder without holding on to anything.

One night, however, my lucky stars completely failed me. As I was descending the ladder with the shutter held high aloft the bottom rung of the homemade ladder broke and I plunged the remaining short distance to the ground where I stopped with quite a jolt. The shutter, which I had abandoned in mid-air, descended with a burst of brand new stars right on the bridge of my nose and from that night on that rather ample member has been quite a little out of line.

Following this fiasco I never tried the balancing act again. I even left the shutter off completely whenever I felt certain of clear skies. Unfortunately, some of my weather predictions were not too accurate and on a number of occasions I was forced to hastily terminate my social activities in a town several miles away and drive madly home to cover up the dome before everything was soaked. My sister Dorothy, always ready with a verse for every occasion, once commemorated these curtailed capers with a long narrative poem of which I can now only recall the following few lines:

> Our hero felt the wind's first lash.
> He saw the distant lightning's flash.
> "My dear," he cried, "I shall return
> Some other night with lips that burn.
> Right now I've got to dash for home,
> For I forgot to close the dome."

We were immensely proud of the new observatory. At that time its nearest neighbor was the original Perkins Observatory, which housed, I believe, a 9-inch refractor and was located on William Street, in Delaware, Ohio, some ninety miles away. Many passers-by stopped to inquire just what our strange appearing building was for. The most frequent conjecture was that we had gone into the chicken business and this was some newfangled kind of brooder house.

It was a great success. No longer was I troubled by dew and

frost collecting on my lens. No longer did the stars in my telescope quiver and dance to the whistling of the wind. I could now observe with more efficiency and in greater comfort and it was indeed a pleasure to have a smooth and solid floor beneath my feet. Without a doubt the observatory was a forward step.

But every step of progress slips back a little too. I now had lost my common touch with all the other denizens of night and only faintly could I see and hear them. Seldom could I note the meteors as they streaked across the sky and never again did fireflies pause before my lens and dazzle me with bursts of starshell fire. Nocturnal bird calls seldom came to me and the pulsing of the cricket's chirp was gone. I also missed an old companion of my early morning watches—the five o'clock whistle that blew each morning at the railroad shops in town four miles away. Last but not least, my gallery of cows deserted me completely.

15 The Comet Seeker

SCARCELY WAS THE 4-INCH COMFORTABLY SETTLED IN ITS NEW home before it had to leave. I received a letter from Henry Norris Russell, the director of Princeton Observatory, offering me the loan of a 6-inch refractor that they were not using. I accepted without any hesitation for a 6-inch has more than twice the light grasp of a 4-inch.

A few days later I came down to earth with a most disturbing doubt. Would a 6-inch telescope fit in my 9-foot dome? According to the figures I came up with it would be a rather tight squeeze if the instrument had a normal focal length, which is close to 8 feet. This would allow less than a foot of room for a size 7⅜ head for when I observe stars near the horizon my head also is inside the dome. I might be forced to cut a section off the top of the pier and give up the stars near the horizon. I could only wait and see.

In about two weeks I got a card from the express office announcing receipt of the shipment from Princeton. I drove to town at once and found three wooden boxes marked for me. After sizing these boxes up with my eye I asked the express agent if there was not still another box in the shipment—one which should be about eight feet long. He looked up the bill but it clearly specified that there were only three boxes. Completely puzzled, I loaded them into the car and drove home.

The first box I opened cleared up the mystery and raised my drooping spirits to an all-time high for when I removed the strapping and lifted the lid, there before me, a rhapsody in dark mahogany and gleaming brass, lay my new telescope, just four feet long—one foot shorter than the 4-inch! I would not require a head-shrinker's services after all.

It was with mixed feelings of regret and anticipation that I removed the 4-inch telescope from the position of eminence it had worked so diligently to attain and then had held so briefly. It had been my faithful companion on hundreds of starry nights and I was fond of it. Its images were sharp and its colors were true. Its out-of-focus rings were a joy to behold. It had served its purpose well, but now I had a scope whose field was twice as wide and that collected more than twice the light. So I shipped the 4-inch back to Cambridge, with the same vaguely guilty feeling that we get when we trade in an old and faithful car that has served us long and well and given us, perhaps, some pleasant memories, and then drive away in a new and shiny model. I never heard from the 4-inch after we parted.

The new telescope was mounted on the pier and once again Vega was invited to attend an inaugural and once again she graciously complied. That opening night happened to be one of exceptional clarity and it was truly a night of revelation for I had never before seen through a short-focus telescope. It was at once quite evident that this telescope was a specialized instrument and its specialty was wide and brilliant star fields rather than high power and magnification. Such regions as the Double Cluster in Perseus, the star fields around Deneb and Gamma Cygni, and the Milky Way in Scutum, Ophiuchus, and Sagittarius were gorgeous quite beyond description.

In the nights that followed I became better acquainted with both the advantages and limitations of the instrument. For variable stars it was superb. It had no finder and it needed none for, with its low-power eyepiece, it was a finder itself, with a field 2 degrees or four full moons wide. It would show stars down to the fourteenth magnitude and occasionally even to the fifteenth, which meant that the majority of my variable stars could now be followed throughout their entire cycle.

The instrument's shortcomings were few and of little consequence since, for my observing, I needed no high magnification. For observing the planets, for separating double stars or for any work which requires high powers and critically sharp images the focal ratio of an objective should be at least f:15, meaning that roughly the length of the scope should be fifteen times the diameter. That of my new scope was only f:8.

The 6-inch was indeed a custom-made instrument. It had been designed with just one purpose in mind: comet seeking. In all probability it had been conceived and built back in the latter part of the nineteenth century, in what might well be described as the Golden Age of Comet-seeking, for it was in this period that such noted hunters as Swift, Tempel, Brooks, and Barnard were vying with one another for each new discovery. The intensity of this competition is quite understandable for, during part of this period an award of two-hundred dollars was made to the discoverer of each unexpected comet. Barnard, then an amateur, is said to have made the payments on his new home in Nashville by means of these tangible trophies of the chase.

I have tried in vain to ferret out some facts about the early life of the 6-inch but thus far no one can tell me when or where it first beheld the light or who its maker was. I did learn that at some time prior to 1907 it was acquired by Princeton Observatory for it was in that year that Zaccheus Daniel, while still a student at the University, used it in the discovery of the bright Daniel's Comet of that year which is so often pictured in astronomical books. In 1909 he picked up two more comets with it, one of these being a short-period comet that often has been seen at subsequent returns.

So actually this telescope that was so new to me was really a patriarch with a long and honored past. But already its birth certificate was lost and the story of its youthful years forgotten. Rather piqued by such neglect, I vowed that some recording of its earlier deeds, as they were known to me, would now be permanently preserved. I whetted up my pocket knife and on the midriff of that wooden tube I deeply carved the name of Daniel and beneath it cut the date of each of his three comet catches.

It seemed to me that if ever human attributes could be invested in a thing of metal, wood, and glass then this ancient instrument now in my keeping must long for one more chance to show what it could do. As to this ability I needed no other proof than the three dates that I had just engraved. I was more than eager to cooperate with it.

But before I could take up comet seeking in a serious way I had to make some changes in the mounting for apparently its last use at Princeton had no connection with comets. It had come to me equipped with a big weight-driven driving clock. This would only be in the way when comet seeking and therefore had to come off. Next, the equatorial mounting had to be changed to an alt-azimuth, for with the former it is quite impossible to sweep across the region around the pole and I wanted no restricted areas—especially in the north.

An alt-azimuth mounting, as its name indicates, has two completely independent motions, one in altitude—or from horizon to zenith, the other in azimuth—or parallel with the horizon. Just how to alter this solidly built equatorial presented quite a problem until Gilbert came to my rescue. He was now employed as a draftsman in a local manufacturing company, furthermore his Uncle Phil was a patternmaker and finally he was well acquainted with Mr. Sudmoeller, the proprietor of a small brass foundry. As a result of this fortuitous combination of contacts the design that Gilbert drew up soon resulted in a brass casting which permitted the equatorial head to be rotated a full 360 degrees when the polar axis was lowered to a horizontal position. In principle, at least, I had now reverted to the old grindstone mounting.

I should make it clear just why this change was necessary. In comet seeking one selects a particular sky area and sweeps the telescope horizontally back and forth across this area until it has all been closely inspected for possible comets. Then one moves to another area adjacent to the first and sweeps over it in turn. In my case I would be working in a dome whose open slit would outline the sky area to be searched. So, in order to sweep across that area horizontally, the polar axis must also be horizontal and parallel to the opening.

One of the early pioneers of comet-seeking was a Frenchman named Charles Messier. He found a number of comets but today he is best known, not for his finds but for his disappointments. Comets, especially faint ones, are not very clearly labeled as such, and Messier was always running across things in the sky that looked just like small tailless comets but that really were faint star clusters and nebulae. In order to systematize his work and save himself future trouble he made a catalogue of all these suspicious-looking objects. To each of these he gave his catalogue number along with its location.

First in his catalogue was M1, now also known as the Crab Nebula in Taurus and in a small telescope it looks for all the world just like a small tailless comet. His M31 is the Great Nebula in Andromeda; his M42 is the Orion Nebula. One would say that some of these objects could not possibly be suspected of being comets but we must bear in mind that Messier did all his observing with a telescope two feet long, with a lens two and a half inches in diameter and magnifying only five times—actually less powerful than the average binoculars. His complete catalogue listed a total of 103 of these objects located all over the sky.

No doubt everyone who seriously takes up comet hunting follows somewhat in Messier's footsteps and gradually accumulates his own file of suspicious characters. As my telescope was considerably more powerful than that of Messier, the true nature of many of his objects was readily apparent and my own personal list contained but few of these, but the 6-inch did show a host of faint clusters, distant galaxies, and nebulous objects too faint for Messier's glass but quite suspect in mine. I simply listed these objects in the order of their right ascension and made a freehand sketch of each one as it appeared in the low-power field of the telescope. In this way, whenever I again encountered one of these objects, I had only to refer to my file to know at once if I had seen it before and, if so, then it definitely was not a comet.

A good star atlas such as Norton's or the Skalnate Pleso, *Atlas of the Heavens* is a prime essential for the comet hunter. Not only are their charts necessary in order to plot the position and thus determine the coordinates of right ascension and declination of any find that he may wish to report but these same charts also show the locations of most of the deep-sky objects such as clusters and galaxies which, in many cases, so resemble comets in appearance. Movement of a suspected object relative to nearby fixed stars is the only certain way to identify a faint comet and this movement can usually be detected with certainty in less than half an hour. However, in some cases the comet, when first sighted, may be so distant from the earth that its apparent motion across the field of view will be extremely slow or again it may be located in a region that has no nearby stars to form a conspicuous pattern in which the comet, by its changing alignment, might quickly reveal its true identity by its steady crawl across the sky.

Since every comet in the solar system sweeps about the sun as a focal point it follows that comets are most numerous in the region near the sun. The western sky just after darkness falls and the eastern sky just before dawn have always proved to be the most productive hunting grounds for visual discoveries of comets.

There are several areas of the sky that are extremely troublesome to the comet hunter for the reason that they harbor such an abundance of faint galaxies that the observer might easily spend the greater part of his time merely in establishing their identity. In some parts of Canes Venatici, Coma Berenices, Ursa Major, and particularly in Virgo three or four of these faint galaxies often will be seen in a single field of my low-power eyepiece. In order to save time and also my peace of mind I completely avoid these regions. A beginning comet hunter can get a good idea of what an average comet looks like when first discovered just by sweeping his telescope through the western portion of the constellation Virgo sometimes known as "the field of the nebulae." One particularly cometlike object to be found there is the eleventh magnitude galaxy listed as NGC 4570. An even easier object is M97, known as the Owl Nebula, in Ursa Major. This resembles a tenth magnitude comet. If, on a sparkling, moonless night, such test objects as these can be picked up without difficulty one can be reasonably certain that both the eye and instrument are well adapted for comet seeking. I do not suggest it as a reliable comet test but it has been my own experience that comets may also be distinguished by a certain peculiar quality of their light. To me there is a definite substantial appearance to the light emanating from a comet which is quite different from the evanescent, fugitive shimmer that finally reaches us from the immensely distant cluster or nebula. It is possible that it is this vast difference in distance that makes this distinction in light quality noticeable.

In my perusal of the long and thrilling history of comet seeking I learned that there are no rigid rules to follow; there is no secret formula by which a find is made. In one of my early acquired books on star-gazing, the splendid *Beginner's Star Book,* by Kelvin McKready, there occurs the provocative statement: "There is no weighty reason why any amateur astronomer should not be the discoverer of a comet. The requisites are a telescope of low power, large field, and generous illumination; a good store of pertinacity and patience, and a fair knowledge of the constellations." In my early comet hunting I often recalled this passage and accorded it a lot of dubious reflection. I had a telescope of thrice-proven efficiency, my persistence I thought at least fair to middlin', and I knew the constellations well. In addition I had a fine little observatory, a good location on the farm with a sky quite uncluttered with trees and lights, and, finally my nights were free to do with as I liked.

Just one thing was lacking—and it was three long years in coming—a comet!

16 Friday, The Thirteenth

As a child I once made the amazing discovery that whenever the first day of the month came on a Sunday then we invariably also had a Friday the thirteenth, that same month. As I grew older I continued to take note of these approximately semi-annual occurrences, not because I attached any ominous significance to Friday the thirteenth, but simply because every truly addicted star-gazer is also a confirmed calendar-watcher who shapes his nightly course according to the faces of its moons.

In 1925 the first day of November fell on Sunday and eventually, in accordance with my still unrepealed law, Friday the thirteenth arrived. I spent most of that clear, crisp, autumn day husking corn, then, after the chores were done and the early supper hour over, I donned my heavy mackinaw, my wool cap, and my sheepskin gloves and started for the observatory. The big elm by the gate to the west of the house now was nearly bare of leaves and, looking up, I could see, silhouetted against the sky, the platform of the lookout station I had built among its obliging branches some ten years before. Beyond the trees along the Auglaize River, Jupiter and Venus made a striking pair just above the overturned Milk Dipper in the low southwest. Directly in the south lay Capricornus and the sharp and steady separation of the twin stars that form its Alpha was a tribute to the clearness of the night.

Dropping out of sight a little north of west was middle-aged Arcturus and just below him, lost in darkness, lay the old but ever new strawberry patch whose fruitful runnered rows had once produced a spyglass.

Before me loomed the observatory. Four years had now passed since its building and the only significant change in all that time was that my old red-eyed oil lantern and my little battery light were now permanently retired. Two slender copper wires now stretched all the way from the observatory back to the house, just high enough above the ground to clear the revolving reel of the old McCormack grain binder. Light and power, at long last, had come to the farm.

In anticipation of its coming I had wired the entire house, installed all the lighting fixtures and had everything in readiness for the power company to set in the meter. They did this on an afternoon when Dad and Mother were in town. It was almost dark when they returned and I had everything ready to give them a little surprise. I hid in a small closet in the kitchen right beside the light switch. They came in and put their groceries down on the kitchen table, then I heard the rattle of the tin match box on the wall and just as Mother was about to light the old Aladdin lamp I turned on the big new light in the ceiling directly above them. Almost blinded by the sudden glare there was a long moment of utter silence. Mother recovered first and looked up at the light. "Well, well," she exclaimed, "so Alec finally arrived." "Alec," echoed Dad, "Alec who?" "Alectricity," replied Mother.

Out in the observatory on that night of Friday the thirteenth I quickly turned the dome toward the west. Corona was sinking low and it contained two star attractions that I checked on each clear night—the irregular variable star R Coronae and the old nova T Coronae of 1866. I found R, as usual, barely visible to the naked aye at magnitude 6.1 while T was still unchanged from its normal magnitude of 10.0. Having no immediate pressing problems with variables I began comet hunting in the low strip of sky in the west which soon would sink from sight. I set the dome due west and examined the outlined sky area to a height of halfway to the zenith, then shifted the dome opening to the right and searched the adjacent area. Another shift to the right and this time I had an area that yielded two Messier objects; first, M13, the Great Cluster in Hercules—discovered by the great Halley in 1714, then, just a few sweeps above it, I passed over the fairly bright cluster, M92, in the same constellation. Doubtless each of these had long ago given comet-hunter Messier a momentary thrill and then a later letdown.

Again I shifted the dome to the right until it outlined an area that slightly overlapped the one that I had just examined. Starting at the horizon I slowly worked upward back and forth in horizontal sweeps across that bounded bit of sky. Sweep by sweep I climbed upward through Coronae, pausing ever so briefly as I hailed in passing the patterned landmark of the R Coronae field, and on I moved into the northern end of Bootes.

It was just above the peak of that kite-shaped figure of the Herdsman that the steady cross sweep of my telescope abruptly stopped. A small round fuzzy something was in the center of that sea of stars! A closer, calmer look and I was sure just what that something was, for extending downward from it I could dimly see a slender streak that could only be the tail of the comet I had just discovered!

It took me quite a while to come down to earth again. To deliberately start out on a comet hunt and then to find one in less than twenty minutes was rather an awesome experience for me and it left me deeply thrilled. Such visitors from space were far from being total strangers to me. I had cut my early sky-teeth on two mighty comets back in 1910 and in more recent years I had followed the nightly courses of all those comets reported in the pages of *Popular Astronomy* which came within the reach of my telescopes. Now, for the first time, I had seen a specimen that was still unlabeled.

Or was it? Had I found a new comet or was it only new to me? Had I merely stumbled onto one that might have been spotted elsewhere weeks before? I knew that the professional observatories subscribed to a telegraphic service as their source of spot-news information but this was far too great a luxury for me, and at that time I had never heard of the Harvard Announcement Cards which convey to the more frugal subscriber this same information about three days later than by telegraph. My only news of any sightings in the sky came to me from the pages of a monthly magazine and many things can happen between publication dates.

I knew that I must make absolutely certain of everything and then get off a wire to Harvard College Observatory. First of all I drew a little sketch that showed the comet's exact position in relation to three or four other nearby stars in the same field. Then I plotted its location on my Upton's atlas. This gave me the comet's precise position in the sky. Next in order was a rough estimate of the total brightness of the object. This should be a simple matter for a variable-star observer. It was done by selecting some star of known brightness which, when thrown out of focus to the same apparent size as the comet, would equal it in brightness. In this case a ninth-magnitude star in the nearby field of the variable V Bootis seemed just right.

There remained two more factors I wanted to know before I sent my wire. These were the direction and approximate rate of the comet's motion. Only ten minutes after I had made my sketch

I could tell that the comet had moved among the nearby stars but I waited still another fifteen minutes and now the field was latticed with the distant treetops. But I had seen enough to know that the comet was moving south—and fast! I already had my telegram composed in my mind and now I wrote it down. It read: NINTH MAGNITUDE COMET ONE FIVE TWO FIVE NORTH FORTY FOUR DEGREES RAPID MOTION SOUTH. Once again I checked that all-important position in my atlas—then raced for the house and the telephone.

I tried three times but got no answer from Western Union. I rang once more and at Central's "Number, please," I asked her why I got no reply from the telegraph office. She replied that it closed at 6 o'clock but that emergency telegrams could be sent at night from the signal tower at the Pennsylvania Railway depot. "Will you please connect me with the tower?" I asked. "I'm sorry," she replied, "there is no local telephone connection."

After I hung up the receiver the full import of the operator's words struck me. I would have to TAKE my message to the signal tower—and I had no car! Dad and Mother had it and where they had gone I had not the slightest idea. I pondered the situation for just a moment, then headed for the barn. I still had the old white bicycle of my high school days. It had taken me to Delphos hundreds of times and it would have to do it once again. I got it out of the corner, felt the tires, blew the dust off the saddle and started for town.

I had no light of any kind on the bicycle and there was no moon, but starlight nights are never really dark and I had no trouble keeping on the road. As I passed the cemetery at a lively clip I remembered my winter walk to town through the deep snow of seven years before. That walk had been made in almost perfect quiet but this time I heralded my coming from afar for my rear mud guard had loosened up or lost a bolt and it clattered in complaint at every rough spot in the road. Several times, at farms along the way, my rattle roused the wrath of irate dogs who voiced their protest at my passing but I plowed on full sail and left their anchored barks behind.

I leaned my bicycle against the wall of the depot and climbed the long flight of wooden steps to the little square room at the top of the signal tower. The operator, a middle-aged man wearing a green eyeshade, had his fingers glued to a stuttering telegraph key and didn't even look up when I entered the room. He continued to worry the key for another five minutes then stopped, walked over to a long bank of levers all of which looked exactly alike, selected one and gave it a pull. He then sat down on a stool in front of a small east window and stared intently into the night. Soon a tiny light appeared in the distance. Like an approaching meteor it grew in brilliance and then sound was added to the picture framed in the window—sound that swelled to a deafening roar as a west-bound passenger train slurred its whistle up, then down the scale and shook the little tower till it rattled.

Suddenly the green eyeshade loomed in front of me and a voice came from under it. "You gotta telegram?" I admitted that I had and handed it to him. He read it through twice. "This some sorta code?" he inquired. "Sorta," I replied. He sat down at his instrument, placed my copy before him and began tapping out the message, his fingers bunched together stiffly on the key except for one divorced digit which wavered about like a lone antenna. Finally he stopped, counted out the words, made a brief calculation and told me the cost. I closed the tower door just as he was pulling down another lever and, carefully feeling my way down the steep flight of stairs, I remounted my bicycle and started for home.

As soon as I got out in the country away from the city lights I began to look about at the sky. It was getting late. The comet and all the western stars had now dropped far below the skyline. By now, I thought, it must be getting dark out on the west coast. What would happen to my message? Would Harvard relay it on so that the big scopes in California could pick it up that night? Or would it arrive at Cambridge only to hear in cultured accents: "I say, here's a good one, some chap out in Ohio has just found that comet that was reported about six weeks ago!"

In the east the entire coterie of winter stars was on display. It was the same bright company that I had named while walking home from town that snowbound winter night—the same that I had seen while lying on my couch of sand that summer dawn at Copus Hill. Still watching them I clattered on and now I roused, in reverse order, each farmer's dog along the way. When I reached home the house was dark but as I put the bicycle away I saw that the car, now too late to serve as a comet courier, was back in its stall in the barn.

Heavy clouds moved in the next day and for an entire dismal week I neither saw my comet nor received a confirmation. Finally, on the twentieth the skies cleared and that night I found the speeding comet far to the south of its first position. Next morning a telephone call came from John Wahmhoff, druggist and town historian of Delphos, whose store served as daytime quarters for Western Union. He read me a wire that had just come in from Harvard that confirmed my finding. The same cloudy spell that had plagued me had also prevented their search until the twentieth. It later developed that the comet had also been picked up—on the nineteenth—in Poland, so far south of my discovery position on the thirteenth that for a time there was some doubt if it were the same object, but sky-patrol plates made in South America on two intervening nights firmly established its identity. Thus assured, I went out to the observatory, opened up my pocket knife, and near the far end of the telescope—a foot or two beyond the inscribed trio of Daniel's early finds—I carved the name PELTIER deep in the dark mahogany. Beneath my hand the old tube trembled as just below the name I cut the figures 1925. The comet seeker had come back to life again!

In ensuing years I made other trips up the steep-pitched stairway to the signal tower but never again did my white-wheeled steed help to carry the message to Harvard. Never again has the occasion been so fondly remembered as that clattering ride on the night of Friday the thirteenth.

17 Southwestward Ho!

ONE SUMMER DAY DURING MY MID-TEENS I WAS RUNNING ACROSS our cornfield when my eye was attracted by a tiny bit of color on the ground. I stopped, retraced my steps, and finally located the source. It was a small slender pebble less than an inch in length and made up of three alternating bands of blue and white material. I had never seen anything like it before, which seemed to me rather strange for I had covered our farm pretty thoroughly for years, not only in the everyday routine of farming operations but also in regular hunts for arrowheads. This had once been well-populated Indian country and each year when plowing and after heavy rains some of these striking examples of Early Americana would be brought to the surface. I had accumulated quite a collection of arrowheads but no such colorful pebble as this had ever come to light.

I could find no one who could tell me what my new find was so finally I sent it off to the Geological Survey in Washington, D.C. for identification. In a couple of weeks it came back, together with a letter that told me that it was a specimen of banded chalcedony. On further investigation I learned that chalcedony is a member of the quartz family of minerals—a family that certainly never made the "Four Hundred" list around here for this is a region of sedimentary limestone laid down by the ancient seas that covered it millions of years ago.

So, my solitary pebble was really a stranger in a foreign land; an innocent little waif kidnapped from its ancestral home far to the north and spirited here by the glaciers that covered most of Ohio some thirty thousand years ago. Now flint is also a quartz offspring and lest the validity of my flint arrowheads be questioned I hasten to explain that they too are strangers in these parts. But in their case transportation was provided, not by the slowly marching glaciers, but by the Shawnee Indians who brought the raw material from a natural outcropping known as Flint Ridge, located in east-central Ohio.

Actually the minerals of this flat corn-belt region are just about as varied as the ships of the Swiss Navy, and it would be difficult to find a less propitious locality in which to start collecting. However, over the years, I had managed to accumulate a number of specimens for there were several limestone quarries within range of my bicycle and these occasionally yielded such associated minerals as pyrite, calcite, and fluorite.

As I have remarked earlier—I save things. I still have my little chalcedony pebble and today it rests on a pad of cotton in a tiny plastic box with a closely fitting cover. It is still my most treasured specimen for it sparked an interest that was eventually to lead, by a most devious route, to wedding bells and a long honeymoon sojourn in the great Southwest. To nine glorious months whose days were filled with mountains, rocks, and desert trails, with canyons, caves, and craters, and whose nights were blanketed with a sequined spread of the brightest stars that I had ever seen.

If my little pebble was the starting point of this winding trail to the Southwest then the first turn to the right came back in my early high school days when the route of my white bicycle led me, morning and evening, past a large, square white house at the edge of town. Many times as I rode by this house I would see a small brown-eyed child with long dark curls playing on the lawn. I vaguely knew that she was the daughter of Homer Nihiser, a local beekeeper. After I left high school I no longer traveled this trail but became fully occupied with the affairs of the farm by day and star-gazing accounted for the majority of my nights—though I occasionally arranged other nocturnal activities that served to keep me in touch with the rest of humanity.

Late in 1925 I took a job as stock clerk in a motor truck factory that had just moved to Delphos. Once again I traveled over the road that had taken me to school some eight or ten years before. Once again I passed, morning and evening, the square, white house at the edge of town, and once again, as I drove by, I sometimes caught a glimpse of the beekeeper's dark-eyed daughter. She was now of high school age and I noticed that the long dark curls were gone and that she didn't need them any more. Eventually we came to have a nodding, then a waving, then a horn-tooting acquaintance. Finally one day I saw her near the street and stopped the car. During the conversation that ensued I found that she had never visited an observatory nor seen the stars through a telescope. The stars soon led to picnics and they to movies and then to swimming in the big stone quarry near the farm. All this, however, led quickly to September and Dottie was off to Ohio Wesleyan University and I saw her only on occasional weekends.

In October 1929 came the stock market crash. For the past two years I had been working as a draftsman at the truck factory. It managed to struggle on for a few months longer, then suspended operations completely, and once again I became a full-time farmer. I was fortunate indeed to have this solid security to fall back to in this period of economic gloom which now had settled over the entire country. Here on the farm life went on, basically, much as before. Turtlelike, we simply withdrew a bit into the shelter of our shell and waited.

Though the cheery jingle of a steady income now was missing and Dottie and I didn't take in many movies, yet, when she was home in summer, we still went swimming in the quarry and the river, we still took hikes and hunted fossils and our campfire steaks were just as thick and flavorful as ever. There was even more time now for observing and I greatly extended my list of variable stars and three more new comets were carved into my telescope. But throughout America the tempo of the times dragged slowly. The dominant tone was a dismal dirge of disappointments and defeat. With us it was a quiet interlude that ended on a happy note.

Dottie and I had now known each other for about five years. We had attended the same high school and the same Sunday school. With my parents we had made summer camping trips to the Smoky Mountains, to Washington, D.C., and to the Lake Superior country. We had gone swimming together in the Auglaize in February and our last three New Year's Eves had been spent back in our woods cooking steaks over an open fire. We had many mutual interests and, as an extra bonus, Dottie had started out in life amid the rarest of celestial surroundings—strange and tenuous surroundings such as earth has known but twice in recent centuries. Deep in youthful slumber at the time, I did not learn for 20 years of what had happened in those hours of darkness nor how it was to figure in my future, for Dottie had been born that night in May while the earth was passing through the tail of Halley's comet!

We were married late in November 1933. Immediately after the wedding we left for the Southwest. We had no definite destination and no time schedule. We expected to be gone until spring and perhaps even longer. We were starting out on a glorious adventure and if we had any worries we left them all behind.

Our preparations for the trip had occupied all of our spare time during the two months that had elapsed since first deciding on our future course. We both knew that we would have to rough it most of the time. There would be no stopping at hotels, no entertainments, no luxuries of any kind if we were to stretch our very limited resources to cover the greatest number of days and miles.

We had mutually agreed on the Southwest for a number of reasons. First of all, the climate should be ideal during the winter months. It also was still a rather sparsely populated region with lots of wide open spaces and if we were going to be pioneers we wanted the right surroundings. Still another attraction that drew us toward the Southwest was that both of us, from our early schooldays, had been avid readers of all of the western novels of Zane Grey who, years before, had briefly resided and played baseball in our own home town, and now we longed to see for ourselves his colorful mountains, deserts, and canyons. A more immediate and practical reason, however, was that Dottie had long had an intense interest in all things relating to archaeology. This interest had been aided and abetted by a college course on Breasted's *Ancient Times* and a more recent reading of one of Ann Axtell Morris' books on prehistoric diggings now had her enthusiasm whetted to the point where she looked toward the Southwest as the Promised Land.

This was an attitude that I fully shared for I, too, felt that somewhere out there was the foot of our own particular rainbow and, while we had no thought of finding there a pot of gold, yet we did have hopes of finding other treasure—for the Southwest is the mineralogical Mecca of America. In the brief time after deciding on the trip we had explored a bit into the remote possibility of adding to our travel fund as we went along. I had written to Ward's Natural Science Establishment, of Rochester, New York, and told them of our proposed trip. Ward's is an old reliable concern that purchased minerals, fossils, meteorites, and other science specimens and sold them to museums, schools, and collectors all over the world. They replied to my inquiry with a list of the minerals that they would purchase from us if we could locate them in suitable size and quality. My lonesome little chalcedony pebble seemed to brighten at the prospect of some kindred company.

We left Ohio under the first snow of the winter and headed directly south. The great adventure was under way at last! Each night we camped in some secluded spot along the road, sometimes in a roadside park, sometimes in a lane or farmyard. Each dawn we ate our breakfast of bacon and eggs or pancakes and our own white-clover honey amid new and strange surroundings, and we felt each day become a little bit more balmy as we left the northern cold behind.

A lot of contriving had gone into our traveling equipment. We had a fairly large wall tent with sewed-in floor and insect screens at door and window. This was a veteran of several family camping trips. Our car, also a veteran, was a 1929 Ford sedan. Odd as it seems today, cars of that vintage had no trunk or luggage compartment of any kind other than a shallow cavity beneath the seat, which held the tools. So I built a large trunk of plywood on the rear of the car and also a capacious cupboard that was bolted to the left-hand running board. The door of the cupboard opened down to make a table and inside we kept all our cooking utensils and our provisions. From the scraps of light sheet metal left over from the building of my dome I made a tiny oven which, when placed over one of the burners of our gasoline camp stove, baked perfect pies and rolls.

Our particular pride and joy, however, was our feather bed. This I had made from an old spring cot of the type that holds the occupant suspended on a wire fabric stretched between headboard and footboard by two rows of small coil springs. I permanently bolted the angle-iron headboard to the top of the rear cushion frame; the footboard, when the bed was in use, bolted to the dash of the car and a couple of wing nuts stretched the coil springs to the proper tension. On top of this wire foundation was placed our feather mattress, then our sheets and blankets. Dottie's contribution to our mobile bridal suite was a complete set of curtains.

A typical day on the road would find us up at dawn or shortly after. I would start the camp stove then, while Dottie prepared our breakfast, I would take down the bed—which process required less than five minutes—then check oil and gas for the day's run. By this time breakfast would be ready. Then the dishes would be washed and repacked and we would be on our way. Around noon we would start looking for a good place to pull off the road and cook our lunch and then again, as evening approached, we would try to stop in time to get supper over before dark.

We traveled at a leisurely pace and even though the Southwest was never far from our thoughts we did not hurry to reach it, nor did we shape our course directly toward our goal. Our first objective was the Smoky Mountains. We had already been there on two family vacation trips and this time we wanted to see them all by ourselves. The region had only recently been made a national park and one could travel from one end to the other without finding a motel or public campground. This was quite to our liking for we could camp anywhere in complete comfort. The Smokies did not let us down. They have a distinctive charm found nowhere else and we spent several unforgettable days reveling in the glorious hazy vistas of late autumn seen from the highest mountains of the east.

As still another means of bolstering our travel fund I had, before we started out, contacted an epitaph collector who furnished me with a lengthy list of cemeteries located within a reasonable distance of our proposed route and we made many side trips, usually over unimproved country roads, in order to locate and photograph these quaint and sometimes weird inscriptions. Following one long and fruitless search for one of these concentrated lyrics Dottie remarked that, while she understood that we would be hunting stones on this trip, she had not realized that I meant headstones.

After leaving the Smokies we stopped at Stone Mountain, Georgia, where we found the geology of this immense dark granite monadnock, eight hundred feet high and a mile long, of great interest. The task of carving a pageant of the Confederate Army into the solid face of the mountain, begun by Gutzon Borglum in 1917, had been suspended at the time of our visit. Angling across Georgia and southeastern Alabama we passed into northern Florida and here we turned westward. Near Mobile we got our first view of the Gulf and, in a little boat, we went fishing one misty early morning on Mobile Bay where Dottie caught a strange creature called a stalked gurnard, which resembled something left over from the age of dinosaurs and pterodactyls.

Each clear night since the start of the trip, no matter where we happened to be, I set up the 6-inch telescope and carried on my usual observing program—for I had brought along my atlas and all my variable-star charts. It had been only a very minor problem to adapt the comet seeker to a vagabond career for its stubby four foot length made it an ideal instrument for travel. However, the wooden tube and the heavy iron mounting were entirely too cumbersome to be portable, so I quickly made a tube out of light-gauge sheet metal and to this I attached all the original optical parts, while Leon Campbell, of Harvard Observatory solved the problem of a mounting by sending me an alt-azimuth bearing for which I built a sturdy wooden tripod. I made a special wooden case for the telescope that held it suspended by rubber strips cut from an old inner tube and this reduced road shocks and vibration to a minimum. This case rode on the rear seat of the car while the tripod was rolled up in the center of our tent roll which then was strapped to our right-hand front fender.

The stars had been out on several of our nomad nights thus far but on each occasion we were camped among trees that left me with only a little open sky so that I had to move about from place to place to see my list of stars. It brought back vivid memories of the weighty grindstone mount that I once huffed and puffed about our yard at home. Of the observing during our sojourn in the Smokies let me merely say that it was gorgeous—in the daytime. It is significant that there has been no stampede by the big observatories to secure sites on Clingman's Dome or Mt. Le-Conte. After all, even the Smokies can't have everything.

Our nights along the Gulf, however, were a much more lucid story. Here we drove many extra miles to find camp sites with a good view to the south, for now, all along the southern horizon I was seeing stars that I had never seen before. From my pasture observatory at home there was always a good clear horizon to the south and on good nights I could see the stars well almost to the ground line. Now, in all the region south of this, down to where the sky dipped into the waters of the Gulf, I wandered as a stranger in a foreign land—wandered as I had that summer morning many years before when the alarm clock woke me to view the winter stars which then were still unknown to me. Here, along the shoreline of the Gulf, for the first time, the open sea formed a part of my night horizon. By looking only to the south I almost felt that I was on my tropic isle.

Throughout East Texas, in particular, we were most favorably impressed with the relatively low cost of living. Here, truly, two could live as cheaply as one—back in Ohio. Gasoline here was thirteen cents per gallon which, in our case, amounted to about one-half cent per mile of travel. Roadside stands all along the way offered an overabundance of pecans, oranges, and nectarines and of these we ate, sometimes not wisely but too well, as we drove along. Prime sirloin steaks cost us only twenty-five cents per pound and on one occasion, in preparation for a rather extended side trip, we laid in a full week's store of provisions for a total outlay of only three dollars and fifty cents. We liked Texas and there was lots of it to like.

18 Cave Dwellers

DOTTIE STRUCK PAY DIRT FIRST. WE WERE CAMPED ONE NIGHT IN a farmer's orchard near the little village of Shumla, which is located right where the Rio Grande River, after making the curve of its northern loop, starts its long southeastward meander toward the Gulf of Mexico. The following morning, as we were cooking our breakfast, the farmer's young son—a boy of about thirteen—came by and, in our brief visit with him, he mentioned that we were only about a fifteen minute walk from the Rio Grande and volunteered to show us the way.

We gladly accepted his offer and followed him down a long narrow lane, through a crudely fashioned gate and across a small pasture field. At the far side of the pasture, just beyond a line of low bushes, the ground completely disappeared and we suddenly found ourselves at the very brink of a canyon that was perhaps four hundred feet deep with quite precipitous walls. Far below us were the swift muddy waters of the Rio Grande. Just to see this storied river brought back to us little bits of history that we had long forgotten—Coronado and the fabled Seven Cities of Cibola, Zachary Taylor and the Mexican War, the Texas Rangers, and, more recently, General Pershing and the long "most wanted" but never captured, Pancho Villa.

We had come out at a sharp bend in the river and from this vantage point we had a good view of both canyon walls, upstream and down. From where we stood we could see several large caves in the canyon walls on our side of the river, though there were none on the Mexican side, as it was more broken and irregular. Our boy guide told us that these were old Indian caves and that archaeology students from one of the Texas colleges had done some digging in them. At this point, with Dottie fairly drooling at the sight of those caves, the boy had to leave to do some work at home. We stood there and looked at those beautiful caves, we looked at the canyon walls, we looked at each other. I waited for her to speak. After all, this was her expedition. Finally it came.

"Well, it surely doesn't take a college education just to climb down those rocks. Let's go."

I do not claim that those canyon walls were exactly perpendicular or that a slip or misstep would have plunged us all the way to the river for actually we would probably have bounced about three times on the way down, but I felt that the final result would probably have been equally disastrous. However, like one of the noble Six Hundred, it was not mine to reason why. The line of least resistance seemed to lie to our right where, about three or four hundred yards away, we could see the entrance to a large cave. Angling downward toward the cave there seemed to be a rather vague line of broken rocks interrupted about midway by a huge rock that jutted outward toward the open canyon.

Cautiously we started out along the broken line in the wall. At first we crept along hand in hand in true honeymoon fashion but soon we held a short conference and decided to abandon the touch technique. It was quite pretty but not at all practical, nor could we, at the moment, recall that mountain climbing had been done that way in any of the movies we had seen. Inch by inch we worked our way along until we finally came to the protruding rock that now completely hid both the trail and our destination from view. Again we consulted but now there was not much choice. We would just have to climb up over that rock and pick up our trail again on the other side. So, leaving Dottie at the base of the rock, I slowly worked my way up its steeply sloping face by digging my fingers and the toes of my heavy boots into any seams that I could locate until finally I reached the top.

No mere words of mine can describe my feeling of empty bewilderment when I looked over the rock toward its other side. It had no other side! There was nothing but a sheer drop of perhaps fifty feet to a narrow ledge of rock, and below that there was only the river.

There was nothing we could do but admit defeat and retrace our steps. Dottie was still down at the base of the rock waiting for me to tell her to come ahead. I explained the situation and told her to start on back but she insisted on waiting for me. By cautiously looking over my shoulder from where I lay spread-eagled on the rock I could just see her head, over the bulge of rock, silhouetted against a background of muddy Rio Grande. The picture, at the moment, had but little appeal for me.

Like Napoleon, I found the retreat far more difficult than the advance had been. Flattening myself, leechlike, on the rock to gain all the friction possible I slowly slithered downward with toes and fingers feeling for each friendly bump and crevice. It is said that in moments of great stress strange mental quirks and lucid images are apt to cross one's mind. I found this to be quite true for, all of a sudden, as I inched slowly backward, I clearly saw before me the title cover of a book that had always reposed in unmolested dignity in our bookcase on the farm. The book was, *The Descent of Man,* by Charles Darwin.

The going became somewhat easier after I came over the bulge in the rock so that Dottie could see me and thus help to guide my groping talons until at last I stood beside her. Getting back to the canyon rim was then comparatively easy. For a long time we lay there on the canyon rim flat on our backs, looking up at the bright blue Texas sky, just feeling grateful for the solid ground beneath us. Finally we rose and walked slowly back to camp. It was late afternoon and we remembered that we had not eaten since early morning. During the course of the long, slow supper I recalled my strange sidelight of the day's adventure and told Dottie about my vivid recollection of *The Descent of Man.* She thought this over for a moment, then said,

"We can just be thankful that you didn't see *Paradise Lost* in your vision."

"Why?" I asked.

"Because," she replied, "it tells about the Fall of Man."

Soon after supper we eased our already aching anatomies into the blessed haven of our feather bed, resolving that so far as we were concerned, the college boys could have their mountain climbing.

Next morning we were right back again. A good night's rest can do wonders, but in this case a competent guide could do even more. We had talked to the farmer's son again at breakfast and, on relating our adventure to him, learned that we had been on the wrong track completely. Dottie promptly engaged his services as guide and half an hour later he was leading us down the canyon on an ample trail that was quite invisible from the point where we first had viewed the caves. Even here, however, the boy kept us in constant turmoil all the way by skipping blithely from rock to rock, dashing up to the very edge of a yawning abyss or peering, with utter unconcern, over the brink of cliffs that I would only have approached on hands and knees.

When we finally reached the cave its over-all size astounded us. The entrance was fully one hundred feet long and about twenty-five feet high in the center. In depth it extended perhaps forty feet back into the canyon wall. For Dottie it was a solemn occasion, the fulfillment of a cherished dream. She entered the cave as a pilgrim enters a holy temple. I feel sure that she would have taken off her shoes if they had not been hunting boots that laced all the way up to her knees. Had there been an ancient skeleton lying around I can well imagine that she would have quietly approached its side, then simply and solemnly intoned: "Mr. Cave Man, here we are."

The cave had been occupied for many centuries for the accumulated debris that covered the floor was several feet in depth and the stone ceiling above us was blackened by the smoke from innumerable prehistoric campfires. As our guide had told us the cave had been partially excavated and we could plainly see where several long trenches had been dug in the floor, though none of this work appeared to be very recent. From this we assumed that everything archaeologically important had already been recorded. Had we found the cave untouched we would have left it intact, for we well knew that many times important evidence has been destroyed by amateur explorers and souvenir hunters.

Our only implement for digging was a short-handled trenching tool that we had recently bought in a surplus store in San Antonio. This was a combination tool that had a sharp pick on one side of the head and a broad adz blade on the other. With this we made a short crosscut between two old parallel cuts and in so doing we felt sure that we were despoiling nothing of value.

The dust that we raised in this digging was incredible. Only one of us could work at a time and even then only by breathing through a mask made by tying a wet bandana over nose and mouth. We knew beforehand that this dust can actually be quite dangerous. If too much is inhaled it results in congestion, followed by fever, and, quite often, by pneumonia.

We unearthed, intact, a number of small pottery bowls. Those near the upper levels of the trench were of brownish clay and were painted with black and red designs while those found at a lower level were much more crude in workmanship and without ornamentation of any kind. From deep in the trench we also recovered a few fragments of basketry as well as a number of very curious relics that consisted of two roughly circular flat pieces of pottery bound together with strings of vegetable fiber. We looked with awe on all these deeply buried treasures for they were by far the most ancient artifacts that we had ever found.

Late in the afternoon our boy guide went home to do his chores. Already the sun had dropped behind the opposite canyon wall and it was growing dark within the cave. Loath to leave a spot so filled with ancient days we sat for still another hour at the entrance, reflecting on those long gone times when this same cave had been the center of the lives of those real First Families of America—the early Basket Makers. Abetted by the gathering gloom we pictured in our minds an age-old time when solid rock still formed this cavern's floor, when brown-skinned naked figures moved about a blazing fire that cast dark shadows on the walls, while children played about these very rocks where now we sat, their shouts and laughter filling all the cave and echoing and re-echoing from canyon wall to canyon wall high up above the rushing river.

I wanted to see the stars that evening from the narrow parapet at the mouth of the cave. At best I could have seen only a slender strip of sky above me which widened somewhat toward the southeast—the direction of the river's flow. However, we still vividly recalled our mountain-climbing perils of the previous day and, having no desire to be out on the canyon trail in the darkness, we packed up our pottery bowls and our bits of basketry and started back up the winding path.

When we finally came out on the rim of the canyon we found ourselves standing on top of a stark and primitive world. The sudden dark of the mid-January night was settling all around, but in the east a full moon hung just above the broken, barren skyline. All across the south, beyond the river, as far as the eye could see, lay the miles on mountained miles of Mexico. The canyon, now filled to its moonlit banks with limpid darkness, lapped at our very feet, though I still could see within its murky depths the inky grotto of our cave. Bats now swam in this stygian river and, like jerky flying fishes, often soared in silent spurts above its airy surface.

Wearied by our long climb up the wall we rested on a large flat boulder which stood on the canyon rim at the head of the trail. We had often wished for a time machine that would run in reverse but never had the longing been so great as on this night when we were perched on that rock above the Rio Grande. This boulder guarded the only trail down to the cave. Every man, woman and child who had ever lived in the cave had passed beside our rock. We could have watched a paleolithic pageant covering thousands of years in the life of early man in America without moving from this spot.

The strategic position of our big boulder brought on another line of speculation. Throughout the thousands of years that man had occupied the cave he must have constantly been beset by his enemies—other tribes who coveted his home or his possessions. No matter how primitive these early cave dwellers may have been they still must have taken every possible precaution against a surprise attack. Would not at least one guard be stationed somewhere outside the cave to warn the others of impending danger? What better vantage point for a lookout station than this very rock? From here one could command a view of the only entrance to the cave. Here one could survey the surrounding terrain in all directions. To us it seemed only logical to assume that, during those long centuries when the cave was occupied by man, each night some Stone-Age sentinel sat here on this same spot and looked about on this same unchanging scene.

As we sat there lost in these conjectures on the long ago the clear Texas sky was full of stars. Even with a bright full moon in Gemini many of midwinter's gaudiest sky pictures were hung out on display. We knew that our early watchman also saw these same stars sparkling in his skies. The same Orion glittered in the south, the same misty Pleiades crossed above his head, and in his northern sky the same Big Dipper circled round the pole, even though the polestar then may not have been Polaris. We felt quite sure—though they left no written records—that the dwellers in our cave knew and closely watched the stars. They must have noted that just one star was always fixed in its location in the north and in their wanderings it doubtless served them as a compass. They surely had a name for the Big Dipper which then circled close about the pole and for Cassiopeia, which rose and set far to the north. Orion, Sirius, Scorpio—even the Southern Cross which long ago rose well above the mountains in the south—all must have figured in their folklore. We liked to think that from this lonely lookout our long-gone watchman often sighted on some star to time the silent midnight changing of the guard.

We took one last long look down into the canyon and then, floodlighted by the moon we walked back through the pasture and up the lane to our camp in the orchard. That night the dust of a thousand years, which we had breathed that day, like the curse of the Pharaohs, kept sleep away for hours. We talked long of the Basket-Maker culture—already old when Christendom was new; we reviewed all the many thrilling highlights of the day, but most of all we spoke of the little pottery bowls. They were something tangible. We tried to picture in our minds those shadowy beings whose dark deft fingers had shaped and painted them. We spoke too of the long, long night of dust that the bowls had passed within the cave and of the vastly different world to which we had, that day, awakened them. How long would this day last before another night, another pall of dust, would close about them?

Dottie finally dozed off, but for a long time she tossed and mumbled in her sleep, still digging in a dusty trench in the cave of her dreams.

19 Hyalite Hunting

FIRST ON OUR LIST OF THE MINERALS WE HOPED TO COLLECT FOR Ward's was a quite unpretentious thing called hyalite. We had never seen a specimen of hyalite, we only knew that it was a clear, colorless mineral which occurred as a coating on certain igneous rocks and that it was member of the great opal family. We had always regarded "precious" opal—the aristocrat of the family—as the most beautiful and most colorful of all the gems and the prospect of hobnobbing with even one of the "poor relations" thrilled us immensely.

Mr. R. C. Vance, the mineralogist at Ward's, had told us that he had heard of the hyalite occurrence through a geology professor at the El Paso School of Mines, who had found it while conducting an extensive field trip with his students through the Big Bend region of Texas. Actually, the nearest that Vance could come to pinpointing the hyalite location was that it was "about forty miles south of Marfa." We were yet to learn what a lot of real estate a pinpoint could cover in the state of Texas, but at that time, from the fleecy cloud on which we were riding, such directions seemed quite ample. We would simply drive to Marfa, a town near the southwest Texas border, then turn south for forty miles, get out our pick and collecting bag and start digging hyalite. It was not quite that simple.

We literally blew into Marfa on the crest of a late January "norther" and at first sight the town looked to us like an island in a storm-tossed sea of tumbleweed. We spent the night in camp there in what was unofficially said to have been the lowest temperature ever recorded there—two degrees below zero. This was indeed a chilly reception to what we had envisioned as the desert southwest, and early next morning we were glad to head south. But before we left town we carefully noted the mileage on our speedometer.

One hour later we stopped the car. We were exactly forty miles south of Marfa. We got out of the car and looked around, but our pick and collecting bag were not disturbed, for even a couple of corn-belt prospectors could plainly see that no geology professor had even led his minions through these parts. Perhaps a 4-H Club had once camped hereabouts or the Future Farmers of America may have held a roast beef barbecue here but it definitely was no place for the United Mine Workers. This was Texas cow country!

We got back in the car and drove on south, consoling ourselves with the fact that at least we had run into sunny skies and warmer weather. At the village of Casa Piedras no one had ever heard of hyalite so we continued on southward all the way to Presidio, on the Rio Grande. Local experts here sent us back north again to a small mining town in the lead-silver district of the Chinati Mountains. This drive was our first experience with really rugged mountain travel.

All about this town were mines and mining a-plenty but the region did not have the type of geology we were looking for. We wanted old volcanic terrain that had known an abundance of thermal or hot water activity, for opal is very like quartz in its chemical make-up except for its greater water content. At the village store, however, an old-timer directed us to a locality several miles due east of town which seemed to him to fulfill the requirements that we described.

About a mile from town we came to the little two-room home of George Dawson, the owner of the area we hoped to explore. We found him a most kind and courteous old gentleman who gave us free access to the entire tract and told us where to find both water and a good campsite. From his doorway he pointed out to us in the far distance, a tall isolated peak that we would pass on the way to camp. This peak, he said, he had long ago named Mt. Bob Ingersoll as he had always had a great admiration for Mr. Ingersoll. Whether this liking was for Ingersoll's flowery oratory or for his philosophy we never knew.

For the first five miles or so we followed a dim trail that led through a region of such strange, weird beauty that we found it hard to convince ourselves that we were still on earth. Twisted lava formations, cinder cones, and wide crevasses, half lighted by the setting sun made it a scene so wild and eerie that it seemed we might be wandering through some region on the moon.

We eventually came abreast of Mt. Ingersoll on our left and continuing on in the twilight we soon came to the campsite to which Mr. Dawson had directed us. We found it a lush oasis in the midst of all this unearthly wasteland. It snuggled up against a mountainous wall which formed the eastern boundary of the region. Here a small stream had been dammed to form a tiny lake which furnished water to irrigate this entire desert garden spot.

Ancient cottonwood and willow trees and a thriving apple orchard all attested to the antiquity of the place.

We pitched our tent in a little grove of willow trees right beside the lake and as we were doing so several tame catfish, between two and three feet in length, came up to the bank and watched the whole procedure. Lying on the ground just outside our tent door were two large metates or grinding stones. These handy built-in kitchen appliances of ancient Indian days were almost two feet square and each had a shallow depression hollowed out in the center for grinding and mixing corn meal. Throughout our stay here Dottie was careful to keep our box of Aunt Jemima's Ready Mixed Pancake Flour well out of sight of these two stony-faced old matriarchs. Finally, just to make this campsite completely unbelievable, not fifty feet away was a heated outdoor swimming pool!

After days of dusty driving the prospect of a swim was so appealing that we did not even wait till total darkness fell. The moon was still a narrow golden sickle in the west as Dottie and I emerged from the shelter of the surrounding willow trees and broke the spangled mirror of the pool into a hundred crazy constellations. For the second time in our lives we found ourselves swimming out-of-doors in February. This time it was not a quick, in and out, double-dare dunk in the icy Auglaize but instead a leisurely soaking in a natural pool that was fed by underground hot springs.

The pool was possibly thirty feet square and the water not more than five feet deep but it was more than ample to envelop us completely with a soft caress so soothing that while we were steeping in its luxury our little floating island of Ivory Soap, our only vestige of the outside world, drifted completely out of sight.

We swam and paddled and splashed about for more than an hour until the moon had dropped behind the western mountain wall. Once as I lay floating on my back, a bright meteor flashed across the leaf-bound square of sky above me. For just a fleeting instant it took me back to Copus Hill where I had lain one night on a contoured bed of sand and searched the skies for just one tardy Perseid. Now, far removed in time and distance from Ohio's wooded hills, I watched the winter stars from a willow-bordered pool among the rugged mountains of the Big Bend country. While thus drifting in the dark I made two estimates of variable stars. The first of these was Mira, in the constellation Cetus. Then, turning in my aqueous alt-azimuth toward the north, I made an estimate of Algol, the Demon Star, in Perseus. Oddly enough, these were the first two stars that were recognized as being variable—in 1596 and 1667 respectively. How many times, I wondered, in all the years that followed their discovery, had these stars been observed from such bizarre surroundings.

The next two days were spent in aimless prospecting that yielded nothing but a monumental tiredness at the close of each day which only a thorough soaking in our warm pool could relieve. Actually, this pool had for us a significance far beyond its mere utilitarian use as a convenient bathtub. It served as an indication of the geology of the whole region. It told us that deep down beneath these volcanic rocks subterranean fires still burned. We now felt even more hopeful that somewhere in these tortured hills and gullies our hyalite was hiding.

On the third day we were up at sunrise and after a quick breakfast Dottie packed a light lunch and we drove over to Mt. Ingersoll. From a distance the mountain, in its isolation and in the abruptness with which it rose from the desert floor, seemed to me to greatly resemble the very striking lunar formation, Pico, which I had often watched with the highest powers of my telescope. However, as we approached we could see that a considerable talus slope, created by the weathering of the mountain, surrounded the entire base. Leaving the car in a conspicuous spot we gradually worked our way around this slope, finding, as we slowly progressed, some very colorful specimens of my first love—chalcedony, though here it was not made up of distinct layers of color, as was my cornfield solitaire. These were botryoidal in form, like clusters of grapes, and we found them in many warm shades of brown, red, and orange.

We were roughly on the opposite side of the mountain and perhaps a mile from our starting point when we noticed, high up on the slope, a group of dark lava boulders. As we slowly climbed in their direction we thought that we could occasionally detect an evanescent shiny sparkle on their sunward side. As we approached, the slope became much steeper, the footing more precarious, but we now had but little thought of this for, as we climbed nearer and nearer that shine we saw became a gleam, the gleam became a glitter, and finally, as we stood at last beside the lava boulders, we knew that here a bit of our Southwest rainbow touched the earth. The sparkle in the rocks had been the sunlight reflected from embedded layers of hyalite!

Like all varieties of opal, hyalite is completely amorphous, or without any crystalline structure. Here it occurred in its usual grapelike form resembling clear, glassy drops of gum coating seams and cavities in the rock. Hyalite has no commercial use and is of interest only to museums and collectors—and honeymooners.

We spent a couple of happy days getting out our order and packing it for shipment and also in searching for any other occurrences, then we reluctantly made preparations for our departure. During our stay of more than a week we had seen no other human beings though we knew that someone had been watching us. On one of the evenings when we returned to our camp after a day of prospecting we found that someone had entered the tent in our absence and had placed—very conspicuously on my pillow—a .30 caliber bullet. On first seeing the bullet there I think we both felt exactly as did Robinson Crusoe when he came upon the human footprints in the sand. Like him we had thought our isolation was complete.

We left the bullet on the table when we went prospecting the next day and when we returned that night it was gone. We never solved the mystery, though Dottie, who dabbles in who-dun-its, theorized that perhaps some young Indian, with a flair for the dramatic, had left the bullet as a warning that we were camped too near the ancestral food grinders or that we should cease our nightly ablutions in the sacred swimming pool. Whatever it may have signified we ignored it completely and our scalps remained intact.

We finally broke camp and drove away with a feeling of

deepest regret. In this vast Big Bend wilderness we had made our first strike in minerals; here we had watched the polestar sink to its lowest in our lifetime skies and here my telescope had probed its deepest among the stars below the rim of my Ohio world. We had found a paradise and we could only console ourselves on leaving with the solemn vow that some day we would return. As yet we never have, though the hope is still alive, for there are miles and miles of rocky wilderness that we would like to explore and there are starlight swims that we will not forget. There are nights of weird, unearthly moonlight and happy, sunny days that we both long to live again. Of all the many places where our caravan has rested, we remember most fondly our camp by the Indian Spring. It was a timeless spot.

On our way out we stopped at George Dawson's little cabin to say goodby to him and to thank him for his hospitality but he was not at home. We left a note for him and when we finally got back to Ohio we wrote him a long letter telling him of our trip and of his part in making it a success. May he and his beloved mountain rest in peace.

20 Night on Mt. Locke

TWO YEARS BEFORE DOTTIE AND I WERE MARRIED SHE HAD AC-companied our family on one of our summer camping trips. On this occasion we had gone north through Michigan, ferried across the Straits of Mackinac into the northern peninsula, and then returned home through Wisconsin. This trip marked our first visit to Yerkes Observatory and our first meeting with the Van Biesbroecks.

I had known Dr. Van Biesbroeck through occasional correspondence since 1925 when he wrote me suggesting that I send Yerkes a collect wire regarding any future comet finds that I might make. With favorable skies this direct apprisal would enable them to secure a position measurement and perhaps even a photograph on the same night the discovery was made—a much more prompt procedure than waiting for the relayed Harvard announcement. Van B, as he was familiarly known, was a Belgian by birth and had been trained as a civil engineer. Later he adopted astronomy as a profession and America as a home for he brought his family to this country and joined the staff of Yerkes Observatory. Here he has done much double-star work with the 40-inch Clark—still the world's largest refractor—as well as a great deal of photographic work with the 24-inch reflector, while his engineering skill has enabled him to design and develop a number of accessory instruments.

About a year after our Yerkes trip Van B. returned our campers visit with one of his own. This time he was on a camping expedition and was accompanied by his charming and talented children as well as two staff astronomers, Harold Schwede and W. W. Morgan—who later became director of Yerkes. After dark that evening we all adjourned to my little observatory where the skies cooperated with a brilliant display of all the infinite host of heaven. Our visitors appeared quite favorably impressed with the wide, bright fields of the 6-inch comet seeker, though I feel sure that my little telescope, which would almost fit crosswise in the tube of the giant 40-inch, must have really appeared to them more like a good-sized microscope.

When next our orbits crossed we both were far from home. While Dottie and I were on our hyalite hunt in the Big Bend a letter from Van B. finally caught up with us at Marfa. The letter had been sent to Ohio and then forwarded to us but it had been mailed at Ft. Davis, Texas, only twenty miles to the north! That evening when we found Van B. sitting by the potbellied stove in the lobby of the Limpia Hotel in Ft. Davis his scholarly face was a study in astonishment. I like to think that something like the law of gravitation was working to bring us together this time.

While neither Van B. nor I could qualify as very massive attracting particles, it was the inverse proportion of the square of the distance between us that did the trick.

The following afternoon we picked up Van B. at his hotel and drove the seventeen miles to the summit of Mt. Locke, where the new McDonald Observatory was to be built. As yet there was no improved road up the mountain and much of the grade had to be taken in low gear but we ignored the protests of the car and finally made it to the top. At the time the site for the observatory had been leveled off and the steel reinforcing framework for the polar-axis supports was in place. Even to the eye it was quite evident, from the relatively low north support, that our travels had carried us much to the south of our home latitude, for the angle that a polar axis makes with the horizon is always a precise clue to the latitude of the place—the nearer to the equator the station, the more nearly horizontal is the polar axis. Many years before, as I was adjusting with such care the polar axis of my cow pasture telescope, I was amazed with the realization that all the polar axes all over the world were parallel—whether they pointed straight up as at the poles, whether at 41 degrees as on my white-ash pier at Delphos or whether they lay horizontal down in Equador. McDonald is the most southern of all the large observatories in the United States—being fully 12 degrees south of Yerkes, its parent observatory.

For the past many years there has been a happy blossoming of large telescopes in our arid Southwest. The sites for these observatories have been wisely chosen with year-round clarity and image quality as the essential factors rather than convenience and ready accessibility. In fact, the more barren and remote the site the better is the assurance against the creeping encroachments of civilization. Mt. Wilson Observatory, in Pasadena, is an example of a location chosen without a sufficiently far-sighted view toward the future. Established in 1904, its 100-inch telescope was completed in 1918 and for more than thirty years was the largest reflector in the world. Today Mt. Wilson is hopelessly trapped by the advancing legions of light.

A few years ago the 69-inch Perkins Observatory instrument was transplanted from an unfavorable location in central Ohio to Flagstaff, Arizona and the move was most successful. A similar retreat seems to be the only real answer to Mt. Wilson's problem of the multiplying millions. One can only hope that before Palomar is similarly engulfed that mankind will find the very simple solution for the numbers racket.

The latest large observatory to settle in this land of starlight and sunshine is the Kitt Peak National Observatory, which has come to rest on a two-hundred-acre mountaintop on the vast Papago Indian Reservation in southern Arizona. It is difficult to imagine any invasion of progress and population ever arising here. The peaceful Papago Indians who own the mountain site have changed but little in the past 300 years. Many of them still believe the earth to be flat and spirits still dwell in their mountains. May their Kitt Peak contacts never change them. Their flat earth is a happy, peaceful earth and there is no gain in just a switch in superstitions.

This move to darker skies is not confined to the giant telescopes alone. The amateur observer whose realm of interest is centered in the depths of space has long realized that a dark transparent sky is of far greater import than is large aperture of scope. More and more these serious enthusiasts are seeking isolated spots in which to carry on their work. During our brief visit in Fort Worth, Oscar Monnig and some of the Texas Observers took us one night to a lonely spot far out in the country many miles beyond the city's glow. Here, on the barren treeless plain they had erected a small shack as their headquarters and here they came on starry nights, singly or in groups, to pursue their varied programs unhampered by a single light or tree or building.

Within the past few months a most ambitious project has come to my attention and this too, like Kitt Peak and Mt. Locke, has wisely headed for the hills. Already a brand new observatory, capped with a twenty-foot dome rests upon a mountaintop in Southern California. Known as Ford Observatory from its donor Clinton Ford—long-time top observer and secretary of the

AAVSO—it will house, as its main instrument, the 18-inch Carpenter reflector built by Claude Carpenter, a one-time native of Wayne, Michigan, and a frequent visitor of mine in Auglaize River days. Under the supervision of Tommy Cragg, Mt. Wilson astronomer, along with Larry Bornhurst and Ernie Lorenz, the entire installation was completed in August 1965. Here, at a site said to be unsurpassed by any in the United States, faint variable stars down to magnitude 18 will be closely studied and programs in photometry and solar activities will be carried out as well.

The 7500-foot peak—higher than Mt. Wilson, Mt. Locke, Kitt Peak, or Palomar—is about 30 miles east of Mt. Wilson and is a part of the San Gabriel Range. It lies in a large government-owned area which insures it against any commercial encroachments in the future. To the north and east of the mountain stretches out the vast Mojave Desert which makes a most effective barrier in that direction.

Having lived my entire life in a region flatter than a pancake my spirits soared to lofty heights when the officials of the project graciously announced that the mountain had been christened Mt. Peltier.

On Mt. Locke that night, as evening approached, the construction workers all went home and the three of us were left alone on the mountain. We made a small fire to heat coffee for the picnic supper we had brought along but before we had finished eating Van B. got up and walked over to a point where he could see the setting sun. He stood watching for a minute or two, then turned and called to us to hurry over and look for the "green flash." We arrived just in time and watched until the day's last beam, a momentary verdant spark, slipped past a distant intervening mountainside. The green flash, the final colored ray of sunlight before extinction, is an interesting phenomenon which Van B. told us had often been seen from the mountain. It is often seen at sea for it requires a distant smooth horizon. From where we stood the sun, as that particular time of year, disappeared behind the inclined slope of a far-off mountain to the west. This inclination was supposed to make the sun's extinction a more gradual process than a horizontal surface would have done, thereby prolonging, ever so slightly, the duration of this brief ray of green. This same purpose is served if the observer watches the flash while running uphill. Later on that evening we also watched the setting of Venus, but from her we got no hint of any green goodby, perhaps because the planet was in its crescent phase and the disappearance too gradual.

I had brought the 6-inch telescope along with me that evening for I wanted the thrill of seeing the stars from a mountaintop. The night was moonless and cold and a host of twinkleless stars filled the Texas sky all the way to the horizon. On several occasions throughout West Texas I had experienced nights of amazing clarity but none like this, and it made my very best Ohio skies seem almost murky by comparison. The difference, of course, was mostly one of altitude for here on Mt. Locke, with its elevation of 7,000 feet, I saw the stars as from a point more than a mile above my home in Delphos.

In the south stood great Orion, higher in the sky by a wide Big Dipper bowl than I had ever seen it from Ohio. Directly beneath it, skirting the far-off mountain peaks, mighty Canopus made its brief arc across the southern sky. Until this trip I had never seen Canopus, the second brightest of all the stars, for it is only in the southernmost states that it is ever seen from continental United States, and even here but briefly. The fact that it was seen at all brought home to me another reason why so many mighty telescopes were settling in the South. The region around the south celestial pole will always be a blind spot for observers in the Northern Hemisphere for it is hidden by the curving bulk of the earth itself. But the farther south we go, the smaller this blind spot becomes. From where I stood on the summit of Mt. Locke I could see a 12 degree wide strip of stars which never could be seen from Wisconsin or Ohio.

The impression that I will longest remember of the night on Mt. Locke had nothing to do with sharp stellar images or the new stars I saw in the south. It was, instead, the feeling I had, while all alone in the darkness, of complete and utter detachment from all the rest of the world. There was absolutely no sound. Earlier that evening Van B. had called this to our attention by asking us to remain perfectly still for a moment and "listen to the silence." We listened in vain for there was nothing up there to make a sound. It was winter and there were no nocturnal bird or insect sounds. There was no hum of wires, no rustle of leaves, no sigh of wind. The mountaintop was a silent world.

Around the horizon earth met sky in a line so vague that it was hard to follow save where the turbulent vapor trail of the Milky Way dropped down behind the earth and in the southwest where a pale cone of zodiacal light still lingered in the sky.

In all this vast expanse the only light came from the stars. I looked in all directions and no distant glow revealed a town's location. Even the thriving town of Alpine, which then boasted of being the largest town in the largest county in the largest state, lay hidden in the dark. From where I stood the whole earth seemed deserted. The skeleton framework of the north support, which rose against the sky behind, might well have been the crumbling vestige of a civilization which now had left the earth.

Suddenly the vision vanished. Both sight and sound came back on Mt. Locke as a crackling blaze appeared not far away where Dottie and Van B. had revived our supper fire to keep warm while awaiting my return. I stowed away the telescope and warmed up for a while. Then we put out the fire and drove on down the mountain.

A gleaming dome now sits upon Mt. Locke, and beneath the dome a mighty 82-inch eye peers out through narrow eyelids into all the nooks and crannies of the universe, while all night long astronomers attend the eye, plying it with patient guidance, coaxing it to show them all the marvels that it sees.

For our final night in Ft. Davis we invited Van B. to have dinner with us in our camp. We spent all of that Saturday morning in getting the camp in shape for the dinner and in getting our equipment in readiness for our departure the following morning. After lunch Dottie went into town to buy the steaks and a few groceries for the dinner. We were right in the heart of the greatest white-face beef country in the world and she came home with three enormous steaks for which she had paid just twenty-five cents per pound.

We still had a good supply of our own homegrown pressure-canned vegetables which we had brought along; corn, lima beans, and peas, and these, together with potatoes, strawberry preserves (homegrown, of course), and an apple pie were, by late afternoon, in various stages of preparation. In a fireplace outside I had a deep bed of coals ready for broiling the steaks and soon these were dispensing their savory smell throughout the countryside.

Dottie had spent many hours in planning this dinner and in collecting various and sundry items essential to its complete success. A perfectionist by nature, she had figured out everything with mathematical precision and now, at last, the final count-down was in progress. At the precise moment I was dispatched uptown to bring back our guest.

Everything had been perfectly timed. When we returned dinner was served. Everything was piping hot and on each plate, as the pièce de résistance of the meal, reposed the most gorgeous steaks I had ever seen. Huge, these were, and thick, each one a juicy, golden-brown portion of pure ambrosia over which Dottie had ministered with loving care and infinite finesse. The sprig of parsley which garnished each steak was itself an epic of design.

A couple of hours later I took Van B. back to his hotel. When I returned to camp Dottie was still sitting quietly at the table. Silently we started in to clear the table but I could see that the bounce was gone from her step, the light was gone from her eye. On one of the dinner plates there still reposed a golden-brown portion of pure ambrosia, untouched, except that the sprig of parsley was gone.

Van B., it seemed, was a vegetarian.

21 The Merry-Go-Round Observatory

DURING OUR SOUTHWESTERN WANDERINGS WE MADE SEVERAL MOST interesting personal contacts. In Fort Worth, Texas we met Oscar Monnig, a long-time fellow AAVSOer and he and the regional group known as the Texas Observers made our brief stay in that area a most pleasant one. In Tucson we spent several days with Joe and Frannie Meek. Joe was an ardent observer of variable stars whom I had known for several years for, while still a student at the University of Wisconsin, he had twice visited my observatory on the farm. Through him we now met Dr. Edwin Carpenter, director of the Steward Observatory of the University of Arizona, who permitted us to use the big 36-inch reflector on some of our lowly variable stars. A high point of our stay in Tucson was our meeting with the retired director of Steward Observatory, Dr. A. E. Douglass, an astronomer turned archaeologist, whose studies of the sunspot cycle and its effect on plant growth enabled him, through the medium of "talkative tree rings" to accurately date a great many of the prehistoric ruins of the Southwest.

One of the few disappointments of the trip came in El Paso. Here I lectured one afternoon to a high school assembly. The next day, after we had left town, I saw in a newspaper that Ernest Thompson Seton had also lectured to a civic group that same day. I owed a great debt of gratitude to the creator of "Rolf" and "Two Little Savages" which I would like to have expressed to him in person.

Back home again in April the farm claimed us once more for the summer. In midwinter I was employed as designer by the Delphos Bending Company which, as a means of diversification, had decided to manufacture a line of children's toys and juvenile furniture. For the next three years we lived in my uncle's house while he was on a job in another part of the state. This had been my father's ancestral home and during all my youth we called it "Grandpa's house." Though old it was well constructed, comfortable, and roomy. Still in good working order was the old fireplace beside which Grandpa spent his later years. Still on the walls hung several oil paintings that Dad had made while living there and still above a doorway, as in my childhood days, hung Grandpa's Civil War bayonet and an Indian tomahawk.

Outside, all about, lay the old familiar stamping ground of my youth. Along the western edge of the property ran the Auglaize River where I had learned to swim. Here was the long strip of bottom land that made a fenceless pasture for our cows. Here was my high lookout by the Indian graves and, along the far north end beyond the orchard, was the old strawberry patch where I had once toiled mightily to buy a spyglass. The house was actually no farther from my observatory than my own home had been. It was merely on the opposite side of the same pasture field so everything was still very convenient for observing. All this was just too good to last, however, for all too soon my uncle's work was completed and now he and my aunt were moving back, which meant that Dottie and I and our three-month-old son, Stanley, would have to settle somewhere else.

We were fortunate in being able to rent a house in town which was not only conveniently near my work but it also stood quite close to history for its small back yard was bounded on the west by the old Miami and Erie Canal. In earlier days this man-made waterway, dug out with pick, shovel, and wheelbarrow, had been the vital artery that sustained the life of towns and villages and farms all along this western border of the state. For more than half a century the canal remained a symbol of efficient transportation. Its eventual demise was, of course, inevitable but in this case it was hastened by suicide. In its latter years its slow freight brought in many loads of steel rails for the railroads whose tracks were creeping out across the Middle West. The swiftness of the new Iron Horses soon retired the plodding mules that pulled the barges.

The towpath of this ancient course, deserted now for more than sixty years, was broad and clear of trees and shrubs and to this once-busy trail in our back yard we brought my new observatory—a novel structure that I had designed and built during the two months that followed the announcement that a move was necessary.

This new structure was proof that necessity is indeed the mother of invention. I needed a building that would be transportable whenever the occasion might arise, for Dottie and I both knew full well that eventually we would be moving again. Over the years, during many a star-filled night, the thought would come to me that there must be some more relaxed and comfortable way of doing my particular type of observing; some way that would not require standing for hours at a time on an observing ladder. Only under the spur of necessity did I finally do something about it.

The idea came to me while sitting in my office chair. I thought how pleasant it would be to hunt comets and watch variable stars from just such a position. I leaned back and slowly revolved the chair. I did it again and this time the four walls and the ceiling became a dome of stars about me. I tilted back my head and found to my surprise that about 30 degrees of head movement plus eyeball movement would permit me to clearly see the 90 degrees from horizon to zenith. It all seemed so simple. Basically, all that I needed was my revolving chair and a rigidly pivoted telescope that would move up and down through 90 degrees with the tilt of my head. I put a big sheet of paper on the drawing board and started in.

Since my observatory was to be designed for both comfort and efficiency it had to be enclosed. So I drew it up as a tiny room six feet square and five feet high. Thus, instead of a revolving chair it became a room that revolved on four rubber-tired steel wheels turning on a circular wooden track—all of which were stock parts that the company used in the manufacture of a children's merry-go-round. The chair, a single type upholstered automobile seat that I found in an auto junk yard, was mounted on an old piano-stool screw to make it adjustable in height.

The telescope to be used was, of course, the 6-inch comet seeker and to keep it just as light in weight as possible I replaced the heavy wooden tube with the one which I had made out of light-gauge sheet metal to take on our western travels. As this telescope was only four feet long this was a distinct asset as it was held very rigidly in a forked suspension bar which pivoted in bearings on both side walls. By looking in a mirror while holding a pencil to the side of my head I found that the pivotal center in tilting my head was right in line with my ear. Therefore the bearings were so located on the walls that when I was seated in observing position, ears and bearings were all in line and thus the eyepiece always held its same position directly in front of my eye. Two heavy counterweights slightly overbalanced the weight of the telescope so that gravity would raise the telescope to the zenith and I had only to turn a crank to bring it down. Thus I got two motions for the price of one.

Just to be sure of the utmost in comfort I added a few refinements. One of these was an adjustable headrest which, during a long observing session, greatly relieves the muscles of the neck. I built a tiny table to hold my charts and then installed a ruby light to read them by. Finally, and the greatest boon of all, my long-suffering feet were rewarded with an electric hot plate sunk into the floor just beneath them. I could not forget the many times I had run round and round my open station in the pasture just to bring my numbed feet back to life. I may have overdone this solid-comfort angle just a bit for many times since then I have awakened with a start from a catnap in that chair.

I never quite got around to inventing my chart viewer. This was to be a small device attached to the telescope right beside the eyepiece. It would consist of a long strip of 35mm. film on which all of my charts had been copied—four charts to a frame. Turning a crank would bring any desired chart into place where I could view it with my right eye through a magnifier by means of red transmitted light while I watched the telescopic field with my left eye. This device never got beyond the drawing-board stage for by this time I had memorized most of my charts and it would only have been an unnecessary gadget. But I did make an eyepiece turret for my telescope out of a couple of closely fitting soup ladles which is still in use today. It is a parfocal affair so that I never need to touch the focusing screw when changing powers.

After 25 years of constant use I am still completely satisfied with my rotating observatory and there is little about it that I would change were I to build another. To the best of my knowledge it was, and perhaps still is, the only one of its kind. As I have it made its only enemy is wind, for part of the telescope tube projects outside the building. This I could easily remedy by attaching a light metal shield on windy nights but these troublesome occasions come but rarely. For my own particular purpose—using low powers with a short-focus telescope—it is quite ideal but for other uses and other instruments it might be entirely unsuitable. When mounted as I have it such a telescope is strictly a solo instrument. It is decidedly not the observatory for the "star-party" astronomer for any movement of the building, such as getting in and out, loses the object from the field. It is conceivable that two persons, if of no more than medium stature and preferably somewhat undernourished could, under circumstances of extreme urgency sit side by side in my observatory but such total togetherness is not recommended.

In actual practice my observatory is simplicity itself. Turning an auto steering wheel mounted directly in front of me will completely revolve the building and thus point the telescope in azimuth, or horizontal motion. Turning another smaller wheel raises and lowers the instrument in altitude from horizon to zenith. My only movements are the manual motions of manipulating these two controls plus the slight tilting of my head. A large sky area is visible to the observer, which is a necessity in quickly finding one's way about. It is always completely dark inside the building except for the small amount of light from the exposed sky area. As the telescope rises in altitude it carries upward with it a light-tight black curtain so that no stray light can enter below the telescope. In this dark environment the observer's eye quickly becomes dark-adapted and remains so.

In the past, various types of telescope and observatory combinations have been designed and built with an eye for the comfort and well-being of the observer. Virtually all of these have been equatorial mountings and have had the eyepiece fixed in the upper end of the polar axis and this has been located in a warm and comfortable room that also held the controls for setting the telescope. However, there were certain disadvantages to all these instruments. They all required at least one extra reflecting surface, and sometimes two, in order to bring the image in to the fixed eyepiece. Each reflection naturally meant a loss of light and a less perfect image. In addition, these mountings often could survey only a limited portion of the sky.

One such instrument was the 12-inch Gerrish polar telescope

which was in use for many years at Harvard. Here the downward-pointing refractor tube formed the polar axis of the mounting. This axis was supported by two bearings, the lower one resting on a short pier on the south side of a two-story building while the upper bearing was placed just inside the sealed window of a room on the second floor. Thus the eye end of the telescope projected into a warm room where the observer sat in solid comfort looking downward into the eyepiece just as into a tilted microscope. The astronomer never got to see the actual stars while using this device but pointed the telescope to various stars about the sky solely by the use of setting circles of right ascension and declination that were mounted right before him. Howard Eaton, an early secretary of the AAVSO, who often used this instrument on variable stars, once told me that all these manipulations made it seem more like driving a car than using a telescope for it even had foot pedals that were used for clamping the two motions.

The most serious fault of the Gerrish mounting was its greatly restricted field, for it could reach only about 80 degrees of declination—less than half of the visible dome of sky. Most of the northern stars were completely blocked out by the building that housed the observer and the lower southern sky was also inaccessible as the flat mirror, which was cradled in the polar axis just below the 12-inch lens, would not reflect so great an angle into the telescope without distortion. Remembering my own nights out in the pasture station, that open-air flat mirror must also have been an easy victim for every frost and heavy dew.

In 1882, M. Leowy of the Paris Observatory invented what was known as the Coudé or "Elbow" equatorial. Its polar axis, like that of the Gerrish telescope, was mounted with the eyepiece in a comfortable upper-story room. However, about midway in this axis, and at right angles to it, a declination axis was placed and at the point of crossing of these two hollow axes a plane mirror was mounted at an angle of 45 degrees. This declination axis actually formed the tube of the telescope with the objective in the upper end, while just above this lens a rotatable box held another 45 degree plane mirror that caught the light from the stars and sent it through the objective to the mirror in the polar axis, where it was again reflected up through the hollow axis to the eyepiece in the observing room.

This mounting also had its inherent faults. It was a rather complicated and expensive structure, it had not one but two reflecting surfaces to dim its images and finally the sky area around and beneath the pole was still blocked off by the observing room.

Yet another strange device for achieving comfort at the eyepiece was the Hartness turret telescope, named for its inventor, industrialist James Hartness. This unique hybrid of the elbow telescope employed as its polar axis a flattened dome that revolved in an equatorial plane on the north-sloping roof of a small building. The elbow tube of the 10-inch refractor formed the declination axis of the mounting with a 45 degree prism fixed at the right-angle bend of the tube. The eye end of this tube was brought through the dome near its base on a line perpendicular to the theoretical polar axis. This tube was free to swing in a north-south direction and this movement, together with the rotational movement of the dome, permitted pointing to any part of the sky. This mounting has some advantages over the earlier elbow telescope of Leowy for it has but one reflecting surface and it has no blind areas. On the other hand its eyepiece, while indoors, is not rigidly fixed but moves about with the dome.

I was quite proud of my little merry-go-round observatory. Its efficiency and ease of operation surpassed my fondest hopes. With memorized star charts two and three estimates of variable stars per minute were routine and before it had been in use a year the new observatory had also proved that it could catch a comet. It would do anything that any alt-azimuth would do—and do it in complete comfort. It had no restricted sky areas but could point with equal ease to any star above the horizon. Since it used no reflecting mirrors its optical system was unchanged and there was no loss of light or definition.

When in 1948 the giant 200-inch telescope on Mt. Palomar

finally swung into action the press made much of the story that, when used at its prime focus, the observer, for the first time in history, would ride with the telescope. I gloated just a little over this, for by then I had been riding with my telescope for eleven years, and furthermore, I had not blocked out a single ray of light while doing it.

22 Records and Recollections

FOR QUITE SOME TIME THE OLD OBSERVATORY ON THE FARM HAD been in need of extensive repairs to both the floor and roof. It had been in active use on nearly all the clear nights from 1921 to 1939 and its deterioration was to be expected, especially in view of the fact that it had been quite impossible to prevent driving rains and snows from blowing in around the base of the dome. Now, with the inaugural of the new and highly satisfactory merry-go-round observatory, we decided to tear the old building down completely and get the ground into use. The lumber in the old walls was still in good condition and this was stored away for later reuse. The dome was placed on the ground back of the barn where it remains to this day, still sound, with its sheet-metal covering only slightly rusted after more than forty years exposure to the elements. The following spring the old pasture was plowed up and soon tall corn hid the spot in the center of the field where seven new comets had come to earth and left their names carved in my wooden tube.

I salvaged a lot of memories from the wrecking of the old observatory and these are still untarnished by time's passing. I had kept detailed records of many of the earlier affairs of the observatory and among these were copious notes on the physical characteristics of all auroral displays seen here over a period of many years, together with references to the sunspot groups which accompanied them. Each night I noted the hour when I began to observe and the hour I retired, also the condition of the seeing and the number of variable stars observed. Other sightings of interest to me, sometimes terrestrial as well as astronomical in nature, were also duly recorded. Unusual meteors, occultations, displays of zodiacal light and gegenschein, and, occasionally, notable storms and snowfalls were chronicled in full.

Admittedly, much of my early recording is of no possible interest to anyone but myself. There also is much that is completely statistical and there are still other entries quite unrelated to the science of astronomy, but to me they all are priceless. In my notebook under the date April 1, 1919, I find the one-line entry: "First butterfly of season seen today—a Pyrameis atalanta." At once the door of memory opens on a vivid recollection of our back-yard cherry tree fairly swarming that spring with Red Admirals. I glance over at my bookcase and my eye soon picks out the green and gold cover of Cragin's *Our Insect Friends and Foes.* I take down the book and find on the flyleaf the date of purchase, March 21, 1919. It was the companion of most of my leisure hours that summer. I recall that I built a walnut display case for my moths and butterflies and then saved my money until

I could buy the sheet-cork lining for the case and the long black-enameled pins for mounting the specimens.

Also from my observatory records I learn that during the month of July 1919 we had an intense local recurrence of Cicada septendecim, the seventeen-year locust, and once again I plainly hear their all-pervading high-pitched whine, so unlike the harsh dismal rasp of the annual species. I find that on March 13, 1920, I saw a brilliant meteor which, even as I watched, divided into twins that side by side streaked all the way across the sky. On January 9, 1921, it is recorded that during a thaw I found numerous cyclops and fairy shrimps in pools of melted snow back in our woods. According to my notes, a heavy wind, one stormy night in March, picked up my entire dome and set it down, unharmed, a hundred feet away, And in among some routine observations I am reminded that I rose on the morning of January 13, 1923, to watch the slender crescent moon blot out the planet Venus from the sky and I still remember the neighbors phoning later that day to ask if I had seen "the star on the moon."

Another treasured artifact that I have lovingly preserved from this paleo-observatory era is the old register book in which many of my visitors inscribed their names. The observatory had literally thousands of visitors, particularly in its early years when it was still a novelty and there was nothing similar in this section of the state. The great majority of these visitors came in the guise of science-class students from various high schools. No doubt some of these students may later have left their footprints on the sands of time but right here may I say that I have found few ordeals more soul-searing that playing host to a class of fifty to seventy-five of these teen-age tornadoes who, at one and the same time, could be underfoot, and on the roof, and ranging far and wide in tender twosomes over all the countryside.

The register holds many names that now are cut in stone. Still other once familiar names I find, and these the lone inscription of the evening. Young, fair, and dewy-eyed these signers were but somehow, through the years, they have contrived, through the subtle alchemy of time, to furnish other broods of signers for my present register.

Like sparkling gems in a modest setting are the occasional signatures of those of the astronomical fraternity who have paused here briefly while en route to some convention or eclipse, or simply on vacation. Among the earliest of these was Van B. with his camping party. A Sunday morning visitor was Donald Menzel, then, I believe, teaching at Ohio State and now director of Harvard College Observatory. In later years came the Boks, Bart and Priscilla, he the director of Mt. Stromlo Observatory; P. Swings of Belgium; Anne Young of Mt. Holyoke Observatory; John S. Hall, director of Lowell Observatory; and Frank Jordan, the director of Allegheny Observatory. A several times visitor was W. A. Hiltner, then still a college student and now director of Yerkes Observatory. On one visit Albert presented us with a little wire-haired terrier which was a faithful friend for many years.

One day in late July 1935, I was working in the cornfield and, on looking up, saw Dad with a silver-haired stranger in tow, approaching through the shoulder-high corn. As they came up I got a closer look and then stopped Dad from making an unnecessary introduction, for there was never another head of hair like that belonging to Clyde Fisher. In addition to presiding over the Department of Astronomy at The American Museum of Natural History and being the curator of Hayden Planetarium, Clyde was intensely interested in nature in all its varied forms. I told him of an unusual find that I had made back near the woods. There, several days before, I had come upon an area devoted to the burrowing activities of a wasp known as Ammophila gryphus. Clyde was not satisfied with merely hearing about my discovery —he had to see it for himself.

Ammophila is a large and likable individual who displays not the slightest concern even when observed, as on that day, by two human monsters hovering in its sky not two feet away. These are predatory wasps who stake out a claim to a small plot of ground and then proceed to dig within its confines a number of burrows. In each of these the female wasp buries a caterpillar which she has anaesthetized by stinging and on its body she deposits a single egg. Then she carefully fills up the burrow. It was the final sealing of these burrows that brought us to our knees in that hot and dusty cornfield—for Ammophila is the insect that uses tools!

When each burrow is completely filled with dirt the wasp selects a small pebble about the size of her head, grasps it firmly in her mandibles and uses it as a tamp to pound the dirt solidly in the hole. Several times that day we watched her perform this amazing feat.

In addition to her tool-using ability the Ammophila wasp has been shown to possess the greatest variety of distinct personal characteristics of any insect. One such wasp might be thorough and painstaking in her work, another might be careless and indolent. One would be calm and efficient, still another would be nervous and excitable. We both owned to a warm feeling of kinship with this insect that had such a wealth of human traits. We thought it significant that the insect in which this advanced ability and personality is so highly developed belongs to the order of Solitary Wasps—an order quite opposed in both habit and culture to the Social Wasps which, like the ants and bees, lead a strictly communal life. Here, we thought, was an example that mankind might well heed. Genius and precocity are qualities of the individual; blind instinct guides the hive and the herd.

While we were so near the woods I told Clyde of another insect that I had watched back there that summer. It was a slender, black, fearsome-looking thing that bore the tongue-twisting name of Pelecinus polyturator and that looked like a flying scorpion, tail and all. It had such a venomous appearance, in fact, that until I found it pictured and described in my big *Insect Book* by Howard, I made no attempt to add it to my collection. We failed to find a single P.p. so we walked back to the house and after a drink of water at the pump, went out to inspect the observatory. Here he photographed the comet-seeker from half a dozen angles and snapped the white-domed building from all directions while our cows supplied a setting of contentment in the background.

Clyde paid us still another visit late that summer. On this occasion I happened to be perched high in the top of our big elm tree taking an aerial photograph of my observatory but even from this bird's-eye viewpoint I again knew Clyde from afar by his flowing silver mane, which this time was in sharp contrast to the shining coal-black tresses of his charming wife, Te Ata. A Chickasaw-Choctaw Indian, the talented Te Ata had been born in a tepee on a government reservation in Oklahoma. She was a gifted interpreter of Indian songs, dances, and legends and it was while appearing at Columbia University that she and Clyde first met. On the farm that afternoon she seemed intensely interested in everything she saw—particularly the leopard frogs in our lily pool and the "witches brooms" in our tall hackberry tree. Following this last meeting Clyde often sent us post cards from various odd corners of the world, for he was always chasing eclipses, meteor craters, and erupting volcanoes and attending Indian pow-wows.

I am sure that Dad missed the old observatory more than I. It had been his idea in the first place. He had conceived and built it and it had gradually grown to be a part of his life for he often took charge of visitor's nights for me and he loved every minute of it. In May 1936 I spotted a faint tailless comet in northern Cepheus which proved to be moving in an orbit that brought it within about fifteen million miles of the earth in late July and August. For about two weeks it was a fairly bright naked-eye object, now with a long filmy tail, high in the eastern evening sky. Many visitors were attracted to the observatory during this time and Dad was a happy host to a group of them nearly every night.

On one of the nights during this rather hectic period I came home after dark and found quite a number of visitors already lined up outside the observatory waiting to see the comet. Dad was inside officiating at the telescope. I had not had a look yet that night and, as comets often undergo rapid changes both in structure and brightness when near perihelion, I took my place at the end of the line. It was a clear and sparkling night, too dark to recognize anyone outside the observatory and darker still beneath the dome.

The line moved slowly, for Dad did his work thoroughly, making sure that each one got a good clear view and also a good clear understanding of what he saw. My turn finally came and as I mounted the steps to look Dad explained how to adjust the focus of the telescope and then recited all the known details about the comet and its discovery. The comet was at its very best that night and, as Dad told me next day, everyone thoroughly appreciated the show—except for the fellow on the end of the line, who took a long, long look and left without even saying, "thank you."

Dad enjoyed showing off Saturn's rings and Jupiter's four bright moons and he became quite proficient in pointing out the constellations. Hercules, though not a bright constellation, was his favorite and he delighted in tracing out the keystone figure in its center. Dad could always spot keystones and squares and compasses without half trying.

23 Sky Visitors

FOR THE DEDICATED WATCHER OF THE SKIES EACH NIGHTFALL IS A time fraught with suspense. Will the skies tonight be just the same or has something happened since I saw them last? Some change may well have taken place during daylight hours or behind last night's covering of clouds that will announce its presence in tonight's clear skies as soon as darkness falls.

I shall always be grateful that my years of watching encompassed the only period in all recorded history that could permit me, on four occasions in a span of only eighteen years, to look upward at the vault of stars and find myself staring, in wide-eyed disbelief, at a bright new stranger in the sky that I had never seen before.

The first of these stellar visitors was that grand spectacle of the century that appeared in 1918 on the evening of that day in June when I had watched the solar eclipse through my 2-inch spyglass on its grindstone mount. It was a remarkable coincidence that this series of four bright new stars began with a total eclipse of the sun. Still more amazing is the fact that it ended on the eve of the next return of that eclipse eighteen years later.

Ever since the days of the early Chaldean astronomers it has been known that eclipses repeat themselves in a period of eighteen years and eleven and one-third days (depending on leap years). Thus the eclipse of June 8, 1918, was a repetition of one that had occurred on May 28, 1900. The duration and latitude of the two were nearly identical. The only conspicuous difference between them being that of longitude and this was due to that one-third of a day in the period. During these eight hours the earth makes one-third of a turn eastward and each eclipse thus falls 120 degrees of longitude to the west of the previous eclipse. While seated at my spyglass waiting for the moon to start that 1918 show I idly figured up just when the next performance of that saros, or series of eclipses, would occur. I added eighteen years, eleven and one-third days to June 8, 1918, and it gave me June 19, 1936. "Gee," I thought, "I'll be just twice as old! I wonder where I'll be in 1936." The swift black shadow of the moon soon lifted from the earth that day and passed out into space. But in some small corner of my mind I filed away a date. I knew that somewhere to my east at its appointed time the majestic saros shadow clock would strike again.

The tardy darkness of midsummer was settling all around as I passed through the gate under the old elm on my way out to my little observatory in the pasture. I looked to the west and the sky was clear as a bell. I hoped it would be just like that tomorrow all the way around the world because tomorrow a total eclipse of the sun would draw a thin black line all the way from Greece, through central Asia, to Japan. Halfway down the dewy trail to the observatory I glanced to the north at the constellation Cepheus. Up there near its northern peak, still hidden from the naked eye, a comet I had found a month before was slowly creeping southward. I walked on down the dim path toward the dome. "If moonlight doesn't interfere," I addressed that blank spot in the sky, "you might become the brightest one since Halley's Comet back in 1910."

I had almost reached the building when suddenly my thoughts came down to earth again. My feet stopped dead in their tracks when I realized just why my eyes had never left that constellation in the north. That figure wasn't right—it held one star too many! And then I saw it. There, just off the eastern boundary, the third-magnitude visitor was standing that had thrown my mental picture out of line.

It was thus that Nova Lacerta, now known as CP Lacertae, first arrived in America. But in Europe and in Asia, where night had come many hours before, the interloper already had been noted by astronomers waiting for the eclipse that would come with the rising sun. Photographs taken on the seventeenth, the night before discovery, showed the star still fainter than magnitude 13. It reached its maximum of magnitude 2 on the twentieth as a slightly paler twin of Deneb in the Northern Cross.

In my observatory on the night of discovery I plotted the position of the nova and carefully estimated its magnitude and then composed two telegrams—one to Harvard, the other to Yerkes. Returning to the house to get my car for another trip to town and the station in the tower by the tracks I passed right by the tree beneath which I had watched, with my spyglass, that eclipse of 1918. Now my question of that afternoon was answered. Now I knew just where I'd be in 1936 when I was twice as old.

During the years between these new stars which so sharply punctuated both the beginning and the end of that saros, two other novae had suddenly appeared. The first of these came in August 1920 and was known as Nova Cygni. It too was first picked up in Europe but it announced itself to me just after midnight on August 22 as I rounded the corner of the woodshed carrying my 4-inch telescope out to the open-air station in the pasture. It attained a bright second-magnitude maximum the following night but its light soon faded. I still observe it once or twice each month and find it very faint and steady at magnitude 16.

In December 1934 Dottie and I were far from home. We camped for two weeks in Hot Springs National Park in the Ouachita Mountains of Arkansas. Here we collected a variety of minerals: quartz crystals, lodestone, variscite, rutile, novaculite, and brookite in the highly mineralized Magnet Cove district. For nearly two weeks the night skies had been overcast but finally it cleared and I set up the telescope. Our camp lay in a little wooded hollow between two high ridges that left the sky open only to the north and west. But on that first night of clearing this was sky a-plenty for suddenly it seemed to be June 8, 1918, all over again as my first glance upward at the stars showed a total stranger of nearly the first magnitude sparkling in the empty void between my old friend Vega and the Head of Draco!

I checked and doubled checked the nova in my atlas, then, though I knew that a visitor as bright as this would already have been seen from many places, I nevertheless composed a wire to send to Harvard. Dottie already was asleep in our bed in the car so I decided to walk the two miles uptown to the telegraph office. Before I had gone three blocks from camp a large empty dump truck clattered past me, headed toward town. I gave chase and when the truck had to stop for a cross street I climbed up on the swinging tail gate at the rear. The streets were rough and hilly but I managed to hang on until we reached midtown Hot Springs where, at a very welcome stop light, I dropped off, hunted up the telegraph office and sent my wire to Cambridge. I was quite content to walk the entire distance back to camp. I watched the new star, now low in the northwest, as I walked along and once, when a dog came out and barked at me, I recalled a clattering bicycle ride nine years before to send a comet telegram.

As I had surmised, my present telegram proved to be quite superfluous for the nova had been found in England about ten days before when of the third magnitude and still rising. It reached its maximum on December 23 when it was almost exactly as bright as its near neighbor Deneb. Thirty years later I still follow this nova—now known as DQ Herculis—at frequent intervals with my telescope and, from time to time, I still find irregularities in its brightness of nearly a full magnitude.

There is nothing that so stirs the astronomical world as the discovery of a bright new star. About a year before the outburst of the very bright Nova Aquilae in 1918, Harlow Shapley, later director of Harvard Observatory, had proposed two distinct classifications for these so-called new stars, namely novae and supernovae. The first named class would include Nova Aquilae and all the other novae that had appeared in our galactic system—with two exceptions. These remarkable two were Tycho's Star of 1572 and Kepler's Star of 1604. These were classed as supernovae, which means that they were at least ten-thousand times brighter than ordinary novae. Both of these amazing stars were members of our galaxy but were located at vast distances from our own solar system. Had either one been no more than a hundred light-years from us it would have shone as bright in our sky as does the full moon! Even far removed as it was, Tycho's Star was said to have been visible in full daylight and was distinctly brighter than the planet Venus. To get some idea of what a spectacle this must have been one should go out some evening when Venus is at her brightest in the western sky and take a long steady look at the planet, then face the north and mentally place her right beside the familiar "W" of Cassiopeia. The term supernova is no exaggeration.

While these two stars are the only supernovae known to have appeared in our own galactic system, several are found each year in other galaxies. At its brightest such a supernova often gives out as much light as the combined light of all the other stars in its galaxy!

Novae are utterly unpredictable. Perhaps it is their suddenness and their element of surprise that have made them so fascinating to me. At any rate, ever since the advent of Nova Aquila, each clear night starts off with a brief once-over to see if there are any strangers in our midst. Oddly enough, our satellite—Echo, makes the finest possible practice nova. Whenever one of these bright artificial satellites is above the horizon one should be able to recognize it instantly simply by the "something's wrong" appearance of that section of the sky.

Early in my acquaintance with novae I was impressed by the fact that after the initial outburst of a nova it gradually faded back to its original brightness. Blissfully ignorant of any of the abstruse astrophysics that might be involved, I wondered why, if it happened once with no apparent ill effects, it might not happen again to that same star, just as a child, unpunished for a temper tantrum, will repeat the performance again and again.

I obtained charts of a number of these old has-beens and began prying into their private lives. Three of these—RS Ophiuchi, T Coronae, and GK Persei still showed irregular variations of at least a magnitude, as though they still were haunted by memories of their former transient glory and longed for one more final fling before they could accept the quiet eons of their golden years.

RS Ophiuchi had apparently been a normal eleventh-magnitude star until 1898 when it suddenly rose to magnitude 7.7, then faded again until it reached its original brightness. I observed this star regularly for twelve years, noting occasional small fluctuations. Then, in August 1933, it actually had its second flare-up, some thirty-five years separating the two maxima. Then, just to prove itself completely undependable, it flared still a third time in July 1958.

My affair with T Coronae was not a happy one. Normally T was a star that hovered around tenth magnitude until back in May 1866 when it suddenly increased until it was equally as bright as nearby Alphecca, the brightest star in the Northern

Crown. In true nova form it soon began to fade and eventually was back as before though with the occasional flutter that seems inherent in the metabolism of some of these stars.

Of all these old novae, T Coronae seemed to me the one most likely to quaff the enchanted herbs of renewal. The star was an easy one to observe regularly as it was located far enough north to be visible at some time nearly every clear night. From 1920 on I watched it closely at every opportunity. For more than twenty-five years I looked in on it from night to night as it tossed and turned in fitful slumber. Then, one night in February 1946 it stirred, slowly opened its eyes, then quickly threw aside the draperies of its couch and rose!

Full eighty years had passed since last the star had shattered the symmetry of the Northern Crown. And where was I, its self-appointed guardian on that once-in-a-lifetime night when it awoke? I was asleep!

I had set the alarm clock for 2:30 A.M. intending to get up and observe some early morning variables. The alarm clock did its part. I looked out the window and the stars were clear and bright, but apparently I was not, for I sneezed once or twice and got the feeling that I was coming down with a cold—or maybe even worse. Self-pity comes easy at 2:30 on a cold February morning so I went back to my warm bed with the comforting thought that I owed it to my family, at least, to take care of my health.

And thus I missed the night of nights in the life of T Coronae. It was the night the spectroscopists long for. It is in those earliest hours of awakening that the newborn star—with all the exuberance of youth, divulges its most intimate secrets.

I alone am to blame for being remiss in my duties, nevertheless, I still have the feeling that T could have shown me more consideration. We had been friends for many years; on thousands of nights I had watched over it as it slept and then, it arose in my hour of weakness as I nodded at my post. I still am watching it but now it is with wary eye. There is no warmth between us any more.

On February 22, 1901, Reverend Thomas Anderson, a Scottish clergyman, was walking home late one night and, in glancing at the well-known constellation Perseus, saw a bright new star of the third magnitude in the very heart of the figure. Dr. Anderson knew the stars and announced it at once as a nova. It continued to increase in brightness until it was brighter than either Capella or Vega. It then faded slowly with many sharp fluctuations, but so gradual was the long decline that it required eleven years to sink to its original thirteenth magnitude. Thus far it has had no recurrence but if restlessness is any criterion, then GK Persei, as it is now known, is a star still searching for the fountain of youth.

Do all old novae eventually recur? Probably not, but no one can be sure, so I still keep a number of these old locations under periodic surveillance. Most of all I would like to see Tycho's Star come back. Its appearance in 1572 was before the telescope had come into use and today we have no means of knowing which one of the faint stars in that region is in reality the sleeping giant. Each clear night throughout the year my first glance is always directed to little Kappa Cassiopeiae to see if, just perhaps, it has a strange companion.

One night last summer, for just a split second, I was sure that Tycho's Star was back, for there, right where it should have been, a glittering object shone, almost as bright as Vega. Well, at least I saw a sight that night that Tycho never dreamed of, for as I watched that bright "star" moved. It was my old friend Echo!

In direct contrast to such transient visitors as the novae we also called, while on our Southwest honeymoon, at the final resting place of an old sky visitor who had come to stay. While we were in Fort Worth our friend, Oscar Monnig, who is not only a long-time AAVSOer but also an experienced meteor observer, drew us a map showing how to find the Odessa Meteor Crater which lay about three hundred miles further on along our route in West Texas. Without his map the crater would have been difficult to find for it is merely a saucer-shaped depression about fifteen feet deep in the center and roughly five hundred feet in diameter and it has no distinguishing rampart of rocks thrown up around its outer rim. This crater was discovered in 1921, only thirteen years before the time of our visit and at that time only a very few fragments of the meteor had as yet been found. Had we known of this crater before leaving home I would certainly have contrived a magnetic device similar to those used by utility companies in locating underground water pipes and gas mains. I had often used one of these contrivances and am certain that we could have achieved some measure of success, for in later years this same region was explored with a magnetic locater and more than 1,500 metallic fragments collected from near the surface. We found a number of mesquite trees of good size growing in the floor of the depression but these really sprouted only yesterday in the actual life of the crater for bones of animals now extinct have been found in excavations made in recent years beneath the floor.

On our homeward leg of this same trip we stopped briefly at the famous Meteor Crater near Winslow, Arizona, where, thousands of years ago, a mighty visitor from space blasted out a hole in the desert rock four-fifths of a mile across and five hundred feet deep. At the time of our visit the entire area was deserted. There was no museum or any other buildings on the crater rim as there are today but in other museums throughout the Southwest we already had seen many specimens of the meteorites which had been found about the rim of this huge cavity. At Stewart Observatory in Tucson we even saw one doing earthly duty as a door stop and we thought it a most menial end for such a bright career. Dottie and I climbed down to the crater floor where, with a tiny horseshoe magnet which I had bought the day before in a five and ten, I scratched around in the loose soil. Everywhere I tried the magnet picked up tiny particles of iron which I assumed were pulverized fragments of that early visitor from outer space.

The most publicized, by far, of all sky visitors have never, like the novae, been charted, photographed, or analyzed. Nor have they, like the meteors, ever blasted out a crater in the earth or strewn their shattered fragments far and wide. This latest rash of visitors from space left not a single trace of evidence behind.

Were I to chronicle my own little history of the twentieth

century I believe that I would refer to the decade following the halfway mark as the Flying Saucer Fifties. It was a period of mass psychosis, when people wanted desperately to believe that we were not alone—that other eyes were watching us. It was not just a field day but a field decade for the crank and the crackpot, for the newspapers bristled with accounts of actual landings and even personal encounters with the little green men.

I am sorry that I did not keep an accurate count of the times I was called on the telephone by people who "saw something" in the sky. Whenever Venus was a bright evening star there would be calls about "that bright light in the southwest," and many of these callers would insist that it had not been there before. Drifting weather balloons seen shortly after sunset also brought on a flurry of excitement as also did an occasional aurora or a bright meteor.

During these years I made fairly frequent lecture appearances and almost invariably, at the conclusion of my talks, someone would ask if I had ever seen any flying saucers. And just as unfailingly they would register disappointment in my negative reply. This finally reached the point where, at the height of the saucer season, at least one person in every audience had seen something strange and the feeling was growing within me that I was being neglected; that life was passing me by. Then, one night, it happened to me.

It was late October. The hour was near midnight. I had been comet hunting for about two hours low in the north in a region just above the Big Dipper. Several times that evening my telescope had picked up faint, foggy spots of light that looked suspiciously like comets, but each time these proved to be either distant galaxies or star clusters. I was, admittedly, just a little drowsy, for the monotonous sweeping combined with retinal fatigue, if too long continued, seems to have an almost hypnotic effect. Then, suddenly, the built-in alarm system that I have instantly stopped the slow sweep of the instrument and brought me to attention. Right in the center of the field of view was the strangest sight I had ever seen in all my years of comet hunting.

It appeared to be a long straight row of very faint stars—all of them of precisely equal brightness!

My first reaction was one of complete and utter disbelief. A row of stars like that just could not be there. I would have seen them many times before, for I was thoroughly familiar with that region of the sky. Nevertheless, there they were, and no blinking of my eyes could make them disappear. I turned on my little red observing light, found a pencil, and got ready to sketch the field in order to fix its location in the sky more accurately.

The real shock came when I looked back into the telescope. That row of tiny stars which I had left right in the center of the field of view was no longer in the center—it was clear over near the edge! But the telescope had been clamped in position and could not have been moved. That row of tiny stars was moving!

I forgot all about sketching. For perhaps half a minute I simply watched, dazed and bewildered, as I struggled for some kind of rational explanation. The stars were moving faster now and seemed to be getting brighter all the time. I dared not leave the eyepiece now for it required constant shifting of both controls to keep the objects in the field of view. It soon became quite apparent that unless they changed direction their line of flight would carry them almost directly overhead. I now was struck by the strong resemblance of this row of lights to a close formation of airplanes. Planes in tight formation do not fly directly behind one another. Each one flies far enough to one side to avoid the turbulence created by the preceding plane. The line of lights that I was watching was in exactly this same formation. To me, this meant just one thing—these flying objects also respected turbulence, so they were not away out in space, but right in our own atmosphere and probably not very far above me!

Even today it is a little painful to recall the thoughts that were going through my head at this point, and I will not dwell on how completely stunned those thoughts and the evidence of my eyes left me, for by now those lights, still in formation, were almost directly above me and bright enough to be visible without a telescope.

Then, slowly, my half stupefied mind became aware that something else was happening. I was hearing sounds from up there! Very faint these sounds were—just the same note repeated over and over at regular intervals. I was still in the observatory, transfixed beside my telescope. I could only see and hear through the opening overhead. In the dark I fumbled for the latch on the door, threw it open and dashed outside. For a moment I heard nothing, then it came again and now so loud and clear that there was not the slightest doubt just what was moving through the darkness right above me. It was a skein of eleven wild geese!

With the sudden letdown of my tensions I found myself shaking with a nervous chill. The whole terrifying thing seemed now so simple as to be absurd. My visitors from outer space were merely Canada geese migrating southward at the approach of winter. I had first sighted them low in the north and had watched them fly across town directly toward me, their light gray bodies reflecting the city lights below.

Ever since that night, each time I watch a flight of geese or hear them crossing in the night, it always makes me wonder. What if that line of flight had not been right in my direction? What if those geese had wheeled about before my eyes and gone off on another course? What if they had crossed the town from right to left and, without approaching me, had slowly faded out of sight? Would I have ever figured out their true identity, or would I, from that moment on have been a firm believer?

I feel quite certain that I can answer this with a most emphatic NO, for I am not prone to accept as true anything that does not make sense or that does not permit minute inspection. Were a flying saucer to land right in my own back yard I still would want to feel it with my hands, kick it with my foot, peck at it with my geologist's hammer, and finally I would want to open up the door and have a look at the occupants inside and those occupants would have to be different from Homo sapiens before I would admit that flying saucers are for real.

In all that saucer decade no one came up with a logical reason why these super beings from outer space made no attempt to communicate with us. The protagonists had their explanation, of course. The spacemen were simply waiting around until we became intelligent enough to talk to them. By this same line of reasoning Columbus would still be anchored off San Salvador waiting for the natives to put on some clothes and become civilized enough for him to land and discover America.

While I will be the first to admit that there are undoubtedly many more things in heaven and earth than are dreamt of in our philosophy, I also will be the last to subscribe to a philosophy that throws logic to the winds. If a belief in flying saucers is something that one wants and needs how easy it becomes to see them and how little one requires as proof of their existence. How many of our other cherished beliefs, our comfortable, soothing, inherited beliefs are also made of this same wishfulness? Were we to examine, critically and impartially, our strange collection of creeds, doctrines, and philosophies how much would we find of truth and reason? How much of fear and superstition? Were we to analyze that creed of ours that came to us attached to our certificate of birth, would we not find that it all resolves into just one basic moral law? A law sometimes quite lost in all the vast confusing structure that centuries of human fears and foibles have built around it.

Sometimes I like to imagine that flying saucers are real, after all, and that some day ships from all the far reaches of the galaxy will converge and silently slide down to earth to visit us. As I picture it, when communication is once established with all these visitors from space, we will learn that each one of them has his own philosophies, his own beliefs, and his own God. And as each describes for us his own particular deity we will note that, in every case, the speaker gives himself a very earthlike pat on the back, for the God of which he speaks is but the speaker magnified.

When finally the conclave ends and each strange visitor streaks for home we will recall that on just one simple truth they all were in complete accord. The Golden Rule alone had needed neither explanation nor defense. It had been as universal as the evolution that had molded their forms and the gravitation that had brought them down to earth.

24 Country Life in Town

In 1940 the owners of the house on the bank of the old canal decided to move into it themselves so for us there came a second exodus. Again we were fortunate, for this time we found a somewhat larger house near the western edge of town. Here, owing to the prevailing westerly winds, the air was a bit less hazy and the seeing considerably better and here too a fairly large treeless back yard provided an even more spacious site for my peregrinating observatory than had the old towpath. But, even as we moved into the new home we knew that it would only be a matter of time until we would move again. Our new surroundings, pleasant though they were, afforded neither the open space nor the quiet solitude that we both deemed essential attributes to the permanent abode that we envisioned.

Dottie and I had always liked living in the country. Her home had been located at the extreme edge of town and I had always lived on the farm. We had spent months of gypsy roaming in the Southwest and, returning, we had lived four untrammeled carefree years in my uncle's house by the river, so it is little wonder that neither of us took too kindly to the confines of a city lot. For several years we tried to be contented with our lot and that we succeeded, in a measure, was only because these were busy, crowded years. Here Gordon, our second son, was born and that same year Stanley was enrolled in kindergarten. We were variously involved in church affairs, in Eastern Star, in garden clubs, and in a devious maze of Cub Scout work. Throughout these years the starlight nights were busy too. Four new comets, dodging treetops, roofs, and smoking chimneys, found their way to the merry-go-round observatory which each clear night pirouetted in the dark of our back yard. Some of these finds whose dates are carved into my scope are listed in the records with hyphenated names which show that somewhere on the wide expanse of earth some other watcher, at nearly the same time, had also seen that comet.

Late one afternoon in the fall of 1948 Dottie called me at the office. In a voice bubbling with excitement she told me that she had just learned that the old Moennig property at the end of Cleveland Street was being offered for immediate sale and asked if I would meet her and the real-estate agent out there right away. When I drove up the twisting drive about ten minutes later, Dottie and Bill Jones, the agent, had just arrived. Together we went through the house. To me the seven large rooms—the living room was twenty-two by thirty-six feet—were just barnlike empty rooms, but Dottie, it seemed, could see them with all the furniture in place, with pictures on the walls and with drapes at every window. Feeling very handicapped by my undomesticated vision I soon left Dottie and Bill discussing the plumbing and the heating system and went outside.

The previous owner of the place had been something of an equestrian. Three or four saddle horses had dwelt here and had roamed freely over the entire acreage together with a couple of large Russian wolfhounds who also had the run of the house. Just to the north of the residence I found a long narrow tract enclosed by a white board fence. This had a gravel path ten feet wide around its entire perimeter which had served as an all-weather track and here each day those prancing steeds had been put through their paces. My first glance at this large bare arena showed me that any spot along its north side would make a far better location for my observatory than any I had found since leaving the farm. I walked over to the northwest corner of the track. From this point only the trees around the house to the south and a few tall cottonwoods farther away to the southwest interfered with an otherwise clear horizon. Later I would come back here when the stars were out and determine the precise spot from which to see them best. I now went back to the car and got a tire iron from the tool kit. With it, here and there about the fenced-in area, I dug test holes in the ground. In every one I found the soil to be a deep, rich sandy loam—such a soil as this should grow strawberries as big as teacups! Meanwhile, back at the house, Dottie had also decided that this place was for us and without further ado we signed our names to the documents which made us part owners of the planet earth.

Much of the early history of "the old Moennig place," as it was known, is now lost in antiquity. The last of the Moennigs to live here were the two maiden daughters of old Henry Moennig, the builder of the house. With their passing the property had been acquired as a residence by the local Catholic priest, Dr. John Sassen. After his death in 1940 it was purchased by Harold Wolfe, vice-president of the reorganized truck company which, in the nineteen-twenties, had employed me as stock clerk and draftsman. Mr. Wolfe and his family lived here a number of years and then sold out to Collin Doyle of the same firm and it was from this keeper of the horses and the wolfhounds that we acquired "Brookhaven," the name that we found emblazoned in large letters above the radio-controlled front gate.

The property was about eight acres in extent and located at the extreme western limits of town so once again the winds were in our favor—any smoke or dust from the town would blow the other way. By great good fortune a four acre parcel of pasture and farm ground that bordered us on the north soon came up for sale to settle another estate and we were able to acquire this also. Now, with twelve acres surrounding us we once again began to enjoy the freedom of the wide open spaces to which we had been accustomed in our earlier years.

Our tract in its entirety is long and quite irregular in shape, with its entire western edge fronting on a small creek which overflows each spring and converts about half our acreage into a temporary lake. It is a sort of split-level property, for the ancient river bank runs irregularly through the center and divides it into two nearly equal parts. The south end of the tract is occupied by a two-acre bit of woods, while the north end is bounded by a railroad.

The large, white frame house sits well back from the street on the brow of the hill that slopes down to the creek. It had been built more than a hundred years ago and had been well designed and "firmly builded with rafters of oak." All of its traditional Victorian "gingerbread" around the gables and porches had been preserved in good condition. The ceilings of the rooms on the lower floor are nearly eleven feet high, the front windows are fitted with etched-glass sashes and the doors still have their antique mirror door knobs. Originally each room had its own fireplace but now the house has central heating.

Ordinarily, when an old place like this changes hands, the new owners start right in making alterations and additions in order that the place will reflect their own conception of modern good living. It was not so with us. We made few changes and even these few were mostly in retrograde motion. What we found here was sound and good. It was a leftover bit of an earlier America and we had no desire to bring it up to date.

The house is surrounded by old pine and spruce trees quite in scale with the high two-story building. We have removed several fences that had divided the property into smaller plots and this slight change has resulted in greater apparent extent and it also now requires considerably less maintenance. We gardened on a rather ambitious scale for the first few years we were here but now we have settled down to one small utility garden, two sizable rock gardens, and a long border planting of perennials and shrubs.

We have endeavored, over the years, to assist Nature in developing the place into a sort of natural sanctuary, particularly a sanctuary for birds, wild flowers, and trees. We very soon found that nature was far more in need of restraint than encouragement for, by the late summer of our first year, all of our lowland area had turned into one vast billowing sea of goldenrod, wild aster, horse nettles, and giant ragweed. Now, while we realize that such extrovert weeds as these may have their place in a wildlife sanctuary by providing seeds for winter birds, some of them also have some very irritating habits. In addition they have other undesirable qualities; their bare, brown stalks are an eyesore for fully half the year, and their roots do not form a solid mat to control erosion. Each springtime flood would make deep gullies where they failed to hold their ground.

In place of these prize itch, ouch, and sneeze producers we now have established a heavy bluegrass turf that never washes out. And to appease the birds for the loss of their winter weed seed we have planted a multiflora rose hedge and another of barberry. These hedges not only furnish an abundance of winter food, they also are excellent cover and, being impenetrable, they greatly discourage the inroads of the sanctuary's Public Enemy Number One, the small boy with the air rifle. On one occasion I tolerated the intrusion of a couple of these neophyte nimrods on their solemn assurance that they hunted only starlings. But when an examination of their booty disclosed a scarlet tanager and a golden-crowned kinglet, they and all their bang-bang brethren were banished for all time.

We are well blessed with trees. There are more than forty species represented, many of them of large size. One magnificent white oak standing near the entrance is at least 250 years old and has a trunk diameter of four feet. The creek bank is lined with large willows and cottonwoods, some of the latter more than 12 feet in girth. There were many fine elms here when we first came but already these were showing the occasional branch of withered leaves that told us, all too plainly, that the trees were doomed. There is no salvation for the elms in rural districts such as this. Strict sanitation and thorough and repeated spraying are, at present, the only known deterrents to the Dutch Elm disease but there is no point in cutting down and burning all infected trees and spraying all those that are still sound when, at the same time, every farm woodland for miles around is filled with dead and dying trees. So we simply let them stand, their graceful forms still beautiful against the sky. Then slowly, year by year, they drop their tattered garments piece by piece until at last, to a drum roll of admiring flickers, they dominate the sylvan stage in shameless nudity.

In all this tragic waste we find the interfering hand of man. Sometimes, to just one person can be traced the contamination of a continent. In 1929 a shipment of burl elm logs from Holland arrived at Cleveland, Ohio. Beneath the bark of those logs lurked the fungus of Dutch Elm disease. Until the year 1850 there was not one English sparrow in all America. That was the year that Nicolas Pike brought over the first of several such imports with the vague notion that they would control canker worms. Our canker worms are with us still but I have not seen a bluebird in more than ten years. Just one person decided that America should have every bird mentioned in the works of Shakespeare. Unfortunately, this included the starling. In 1890, fifty pairs were liberated in Central Park. On any autumn evening now I can count more than ten-thousand of these delta-winged demons flying over my home toward their nightly roosts in the maples in the heart of town.

A plague on all these foreign imports! But we have, in part at least, retaliated. For their elm disease we paid them back in kind with American foul brood, the bane of the beekeeper. And for their noisy pests we gave them jazz and crooners.

Ours is a limestone region and we make no attempt to introduce any trees, shrubs, or wild flowers that are not completely at home in our mildly alkaline soil. Exotics of any kind may be a challenge for some but here we can find challenges a-plenty without inviting more. A rhododendron can be a thing of beauty but so is a redbud or a dogwood. A ladyslipper is a lovely wild flower but so is a trillium, a bloodroot, or a hepatica. Mere rarity, in itself, is not a virtue. The edelweiss on the mountain peak is no more beautiful nor desirable than the buttercup down in the valley. To accumulate a large variety of rarities from Venus' flytraps to fringed gentians to lofty alpines could be a fascinating hobby if one had nothing to do but cater to the exacting requirements of each one. But this is not nature's way and it is not for us. Our plants must fight the battle for survival with a minimum of help from us.

When we first moved here we had visions of gradually restoring neatness and order to the entire place, of giving it a sort of well-groomed parklike appearance. These grandiose ideas soon vanished when we came to engage in daily combat with such implacable opponents as the ravages of time, the vastness of our space, and the fecundity of nature. While we would be cutting up a fallen willow tree the grass would be getting too high to mow. While we would be mowing the grass the fences would need painting and as we painted the fence the tent caterpillars would be bivouacking in the apple trees.

Finally we compromised. We held our own small family conference and quickly decided the fate of each of the various components of our little world. The wooded area we would return to Mother Nature. The turf areas we would try to keep free from noxious weeds, and the fences we would gradually replace with barberry hedges which, though not as picturesque as wood, were still traditional and much less demanding of our time.

In spite of the terms of the conference the woodland is more a protectorate than an independent kingdom for we do not hesitate to make a raid across the border to quell an uprising of honey locust seedlings. These we cut down without mercy for they are clearly in league with the Devil. Their soft and delicate foliage is but a snare and a delusion which masks a bristling array of vicious thorns which are a constant menace both to tender flesh and to pneumatic tires as well. We also make certain that any up-and-coming young allies such as mulberry, hawthorn, ash-leaved maple, and wild roses have land grants of their own. We even retain a brushpile or two.

There are few things more dear to the hearts of the smaller, low-nesting birds than an old brush pile. At his home in rural Connecticut, Edwin Way Teale has what probably is the ultimate in brush piles. It is quite a large affair, built just like an igloo, and inside it, surrounded by this latticed dome of brush, the writer-naturalist sits at a rustic desk in comfortable seclusion, spying on the unsuspecting songsters just outside.

On our highest ground, about a hundred yards north of the house, my rotating observatory was placed. Here I had a good horizon, particularly in the west and north. To the east, over the town, the seeing to a height of about 30 degrees is not good in the early evening hours due to smoke and haze, but this usually has cleared up by ten o'clock. In the south the tall pines around the house punctuate the skyline in a few places and through my telescope I often have observed low southern stars in a Christmas-card setting of swaying pine cones. Directly to the south there is a narrow valley of sky between the house and the giant cottonwoods which briefly lets me watch those southern stars that only rise a shallow Big Dipper bowl-depth above the south horizon. Thanks to the cooperation of the Ohio Power Company and my nearby neighbors I have no troublesome lights to contend with. There are a few distant street lights visible in winter but when the trees are in leaf even these are blotted out.

My merry-go-round observatory had now been called upon three times to demonstrate its portability. Here, at last, I felt so sure that I would not be asking it to move again that I cast for it a heavy slab of concrete and to this firm base I bolted, for all time, its round much-traveled track. We finally had found the spot we really liked. We both were glad to settle down.

Then, one night in July 1959 I was called to the telephone. It was a long-distance call and the voice on the other end of the wire was offering me, as a gift, a 12-inch Clark refractor—complete with observatory, transit room, and all the trimmings!

25 The New-Old Observatory

IN THE SAME YEAR THAT DAD AND I BUILT OUR FIRST OBSERVATORY, Miami University, of Oxford, Ohio, also started building one. Their observatory, like mine, had been placed in an open area with a good view of the sky in all directions. But now all this had changed for Miami, long ago the little college where William McGuffey once taught while compiling his celebrated series of McGuffey Readers, was now a large state university and their telescope, which had once looked out, clear eyed, at all the universe had now become myopic and a cataract of taller buildings closed around. Finally, as a last crushing blow, a new dormitory was to be erected on the exact spot now occupied by the observatory.

Astronomy has never been a popular subject in college. Students do not stand in long lines waiting to sign up for it and Miami was no exception to this rule. So the decision was made that the old observatory must go and that for any future courses a small portable instrument, to be located on a flat-topped roof, would suffice. It is, indeed, an ill wind that blows nobody good, and the fact that my older son, Stanley, was a student at Miami and had taken a semester or two of astronomy under Professor Miltenberger, was undoubtedly a guiding factor in the fair breeze that now had blown my way.

We drove down to Miami the day after the surprise phone call in order to get a better idea of the condition of the equipment and to see how it could best be moved. I checked, first of all, the most important item—the 12-inch objective lens. It was quite dirty but seemed to be without a scratch. I could see a few tiny bubbles in the glass and one edge had a small chip that was almost completely hidden by the retaining ring. The driving clock apparently was in working condition and we found one eyepiece of sub-standard diameter. Over against one wall was a well-constructed counterbalanced observing chair and ladder which had been built by Warner and Swasey of Cleveland.

The attached 12 by 20 foot transit room was equipped with a 3-inch Gaertner transit and chronograph and a large Howard sidereal clock. In a wooden case was a micrometer for the 12-inch and this instrument, presumably, was also made by Clark.

The suddenness of all that had taken place in the past twenty-four hours had left me in something of a daze. It was still hard for me to believe that the overnight zoom from a stubby four-foot telescope to a sixteen-foot monster was real. But just to be sure that it was not all a dream, when we finally left for home that night, the 12-inch lens and the lone eyepiece went with us.

Knowing absolutely nothing of the monumental task of moving

a sizable observatory a distance of 125 miles, I told the whole story the following day to Louie Justus, the president of the company that employs me. Louie was intensely interested in the project and it was characteristic of his generous nature that he offered at once to take over all the details of making the move—and it was equally characteristic of me that I accepted his offer immediately. Within the hour he had wheels turning and telephones ringing. He hired a contractor to do the actual work and he enlisted the aid and cooperation of a number of local firms and interested citizens and organizations and to all of these I am eternally grateful.

During all my many years of association with Louie he had always impressed me with his meticulous attention to detail and his insistence on thoroughness and prompt dispatch. In my hour of need it was truly a great relief to be able to place my troubles in his competent and willing hands. But there once had been a day when my faith in him had been something less than absolute.

About three years earlier I had occasion to consult with a moldmaker in Indianapolis concerning a new model that we were bringing out. On previous trips of this nature I had always driven there in my car—for the distance was only 160 miles. But now, as time was of the essence, Louie announced that he would fly me there. In former years he had been a most enthusiastic pilot and had his own plane—a 4-place Beechcraft Bonanza—but of late his fervor had faded considerably and he had not flown for some time and was, in fact, trying to dispose of his plane. For my own part, my boyhood craze for airplanes had never gone beyond that plane that did its tailspins tethered in the branches of the oak tree on the farm, but in the present instance I had no choice but to fly with Louie.

We loaded the two plaster models we were taking along into the plane—one up in front with Louie, while I held the other in the rear seat. With everything secure we taxied to the far end of the runway, turned about and faced west ready for the take-off run. Just at this point Louie idled the engine down, then reached over and took something out of the compartment in the dash and, holding it closely in front of him, he sat and stared at it in complete absorption. After two full minutes of this my curiosity finally got the better of me and, shifting the plaster model onto the seat beside me, I raised up enough to look over Louie's shoulder. What I saw filled me with horrified dismay. Louie was perusing the pages of a small brochure. It was a pamphlet issued by the Beechcraft Company that told, with illustrations, just how to fly that plane!

Early one morning about three weeks after my trip to Miami University to inspect the observatory the contractor called to say that everything was loaded on their trucks ready to start for Delphos. I was waiting for them in my car at the edge of town when they arrived late in the afternoon and led the strange convoy through town to our open field near the railroad. Here we had a large crane, which a neighbor factory had offered us, waiting to lift the cargo off the four trucks onto the ground. The transit room came intact and was set on blocks directly above what was to be its permanent location. The observatory proper had been an octagon and this was delivered to us as eight separate walls. Due to highway restrictions it had been necessary to saw the twenty-two-foot diameter dome into halves and these came nested together on a single truck. When I had seen the observatory at Miami three weeks before I thought of it as a thing of beauty and a joy forever. Here, dismembered, eviscerated, and scattered over the landscape it was a most appalling and discouraging sight.

Long before the observatory arrived I had decided exactly where its permanent site was to be. It was not a decision reached by hundreds of star-image tests on dozens of mountaintops such as were the searches that ended on Palomar, Kitt Peak, and Mt. Locke. I had only two possible sites to consider. One of these was our farm four miles away. The other was right here at home. The farm could offer complete freedom from lights, smoke, and haze as well as a good horizon all around. Its one big drawback was its distance, and I knew that I would use the telescope much more if it were right here beside me. There are usually several nights each month when there is a short interval of good seeing before moonrise in the early evening. With the observatory located at the farm I probably would miss most of these. There are also many other nights when I observe in the early morning hours after the moon has set. I felt quite certain that often I would hesitate to drive out to the farm for an hour of observing, especially if clouds were a possibility. Furthermore with two sons who seemed increasingly dedicated to the proposition that the car should never cool off, it would be well to keep the new observatory within walking distance. Here in town I would be less than two minutes walk from the telescope and, in the event of deep snow, I might drive right to the door and thus even a five-minute clear interval between clouds could be utilized. The seeing, while not as good here in the early evening as on the farm, is nearly equal in the hours from 10 P.M. until dawn.

So, I decided on a location about fifty feet north of my small observatory. This had the advantage that the same power lines could service both buildings. Also, the two instruments could be operated concurrently, so that any doubtful object sighted in the 6-inch could quickly be examined under the much greater power of the 12-inch.

The unique location of the new site in relation to adjoining properties gave me every assurance that the observatory would never again have to suffer the indignity of being hedged in and spied upon by taller, baser structures. On the north we are bounded by a one-track railroad whose trains are pulled by friendly smokeless diesel engines. To the west and southwest, beyond the creek, lies a most effective buffer state, perhaps a hundred acres in extent, occupied by a large stone quarry. To the south stretches the full length of our own property, while to the east are long-time neighbors who cooperate with shielded lights.

The dismantled observatory had arrived in September and for the next two months I was occupied in laying out the exact location for the new foundation and in building concrete forms. Here my lifelong trait of saving things now stood me in good stead for these forms for the new building were made from the lumber salvaged from the wrecking of the old observatory that Dad and I had built nearly forty years before. Several very interesting problems came up during this period of planning. One of these was that of identifying the various wall sections and marking them for their future locations. There had been no opportunity for me to mark any of this material before the building had been torn down and it was now like assembling a gigantic jig-saw puzzle. Another most intriguing operation was carried out early one clear December night. This was the problem of placing, in exactly the correct location, the form for the concrete pier that later would support the equatorial mounting of the telescope. All respectable equatorials point due north or south. So my present task was basically the same one that had confronted me that chilly night out in the pasture when I had wrestled with my white-ash pier. Only my approach was just a little different. First of all I had to establish a meridian, or true north-south line, and this was the reason for working in the gloom of night, for again I made the alignment by the stars themselves, but now the stars I used were in the north—not on the equator.

The true north point in the sky is about two moon diameters from Polaris and almost precisely on a line between that star and the star Alcaid in the handle of the Big Dipper. Shortly after dark, on this particular December night, Polaris, the true celestial pole and the star, Alcaid, would all three be on the meridian. From a ladder on the north side of the observatory site I suspended a long plumb line and from the south side I dropped a much shorter line suspended from my camera tripod. With my flashlight I illuminated both plumb lines, then, at the precise minute, I lay on the ground by the short line and by laterally shifting the tripod I maneuvered so that both plumb lines cut right through both Polaris and Alcaid. A stake was driven in the ground at each plumb bob and a line connecting these stakes became my meridian and I had only to build my pier parallel to this line.

This was really a two-part problem, the second part being the proper placing of the pier in relation to the observatory foundation. I had to be certain that when the telescope was finally placed in position on the pier the pivotal center of motion of the axes of the mounting would be fairly near the center of the dome. I did not want my telescope projecting out through the slit in the dome as I had often seen them pictured in cartoons.

When the telescope came to us from Miami we had it temporarily placed on a couple of heavy timbers and there we gave it a complete coat of red-oxide primer and then covered it with a large sheet of plastic. One clear autumn night our curiosity got the better of us and we decided to have a look through the telescope even though, mounted on the ground as it was, we could point to nothing higher than about 30 degrees above the horizon without the eye end striking the ground. Stanley and I carried out the objective and eyepiece holder from their place of storage in the house and fitted them into the tube. That first night's trial was a bitter disappointment. Every bright star was surrounded by a purple halo!

Stanley thought he recalled seeing these halos when he had used the instrument at Miami. Nevertheless, I was sure that something was wrong for this was a Clark objective. It had been ground by the makers of the world's largest objectives. The 40-inch Yerkes, the 36-inch Lick, and the Washington 26-inch, which first had seen the Martian moons, had all been ground by the Clarks. Furthermore, I once had looked through the 40-inch at Yerkes and there certainly were no such halos around the stars at Williams Bay. I felt like one who had acquired a Stradivarius, and found it made of cottonwood!

We took the objective back to the house and there, in the dining room on our softest rug, in fear and trembling, I removed the lenses from their cell. Sometimes a cloudy film of fungus growth attacks the inner surfaces of old lenses but these, when critically examined, were clear and sparkling.

In my 6-inch telescope the two elements, which together make up its objective, are in actual contact with each other, but in the 12-inch they were very slightly separated by a narrow paper spacer that Dr. Anderson, of Miami, had told me, with a twinkle in his eye, was just the thickness of a one-cent stamp. It was this paper spacer that set me to thinking. It was quite invisible when the lens was in its cell, so the fact that its presence and even its extreme thinness were known, proved that at some time in the past someone had taken this lens apart. Could it, in some way, have been reassembled incorrectly? To the eye the two surfaces of the double-convex front element seemed exactly alike, but just supposing there actually was a difference, too slight for me to detect, in the curves of these two surfaces. This could make a difference in image quality. Carefully I turned the lens over and reassembled it in the cell. This time, however, that face which had so long looked out upon the stars, now snuggled almost cheek to cheek against its concave mate.

Once again we carried the objective out to the telescope and secured it in the tube. The bright star Rigel now was low in the southeastern sky. I carefully centered it in the finder. I shifted my eye to the telescope and, with fingers that were cold and shaking, I turned the focusing screw slowly inward until the star became a blue-white diamond and the tiny gem beside it stood out sharp and clear.

I feel that the three long-departed Clarks must have rested a little more quietly after that night—for the halos had disappeared.

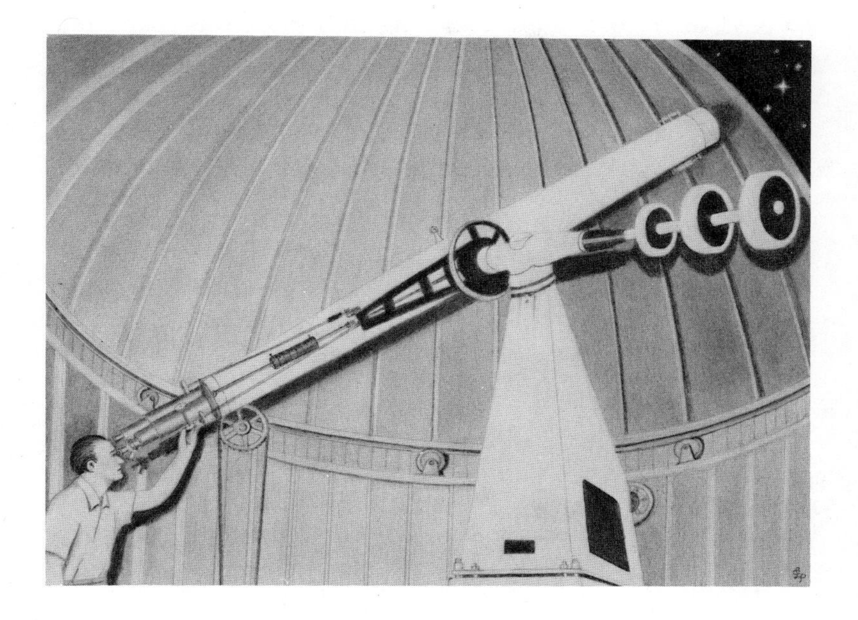

26 The 12-inch Clark

DURING THE MILD DAYS OF FALL AND WINTER, WHILE I WAS BUILDing the forms for the concrete foundation and pier, Russell Brunk, a fellow star-gazer, with hammer, nails, and paint, grafted back together the jagged wound that the saws of the movers had made in the dome. Then, on a warm Saturday in April, a crew of volunteers made up of neighbors and astronomy club members, together with Russell and my sons, Stanley and Gordon, poured the concrete walls while Dottie, a modern Evangeline, bore to us from time to time flagons of home-brewed lemonade.

Came May and with it the building contractors. Neither Ray Ulm nor his assistants, Bob Miller and Paul Grilliot, had ever seen an observatory but in just four days their work was completed. They had erected the walls, mounted the telescope on its pier, set the dome in place and repainted it, built concrete entrance steps and even devised an improved method of operating the dome shutters.

Only in the placing of the dome had any hitch occurred. This dome is a sturdily built affair 22 feet in diameter and 11 feet in depth. It is made of 33 laminated wooden ribs with a covering of ¾ inch boards and an outer skin of painted canvas. Attached to the underside of the circular base ring is the heavy steel rail on which the dome revolves. This base ring also carries the circular cast steel rack which engages the motor pinion to turn the dome.

The big problem was to lift this hemispherical shell weighing several tons to a height of thirteen feet and place it on top of the observatory walls. The borrowed crane that had been used to raise the wall sections into position was hooked to the center of the dome. Then, at Ray's command the cable tightened, the outstretched boom swayed and trembled, the crane reared up behind, then gave a frightened snort and quit. It had neither the capacity nor the length of boom necessary to lift the big dome off the ground and place it on the walls.

"Hold on a minute, boys," yelled Ray, "I've got another idea." He made for his car and drove away in a shower of gravel from the spinning wheels. Bob and Paul busied themselves at other tasks until, half an hour later, we heard an approaching siren and, looking up, beheld a strange procession coming down the street. In the lead was a police prowl car with flasher blinking; behind it rolled an enormous crane with its long boom swaying out in front, while bringing up the rear was Ray.

He had remembered that, a dozen blocks away, a new high school was being built and he had persuaded the contractor to let him rent their mammoth crane for just an hour. The lengthy boom of this colossus was fastened to the dome and again, at Ray's signal the engine gave a mighty roar. Again the crane's rear wheels rose four feet in the air, but the dome, though shaken by the ordeal, still remained upon the ground.

Then Ray had still another bright idea. He quickly backed the small crane up against the rear end of the new arrival and chained the two together. This added weight provided all the ballast needed and this time the dome rose breathlessly aloft just like a giant pasture mushroom and settled slowly on the eight big wheels waiting to receive it.

What had seemed a Herculean task to me they had accomplished with an efficiency and a nonchalance which, had I not known them, would have told me plainly that they had been rebuilding old observatories all their lives. These three all became greatly interested in using the telescope and later on brought their families and friends and even church and school classes out to see the moon and planets. They still continue to do me many favors far beyond the call of duty.

We made a great improvement in the dome-turning mechanism. At Miami this had been done by means of a large hand crank—usually operated by a couple of stalwart students. Since I would generally be working all alone in the observatory, I needed a drive that would rotate the dome in either direction at the turn of a switch. A friend in need heard of this and gave me the special three-quarter horsepower motor which was perfect for the job. Then master mechanic Joyce Everett, together with Russell, designed and installed a fearfully and wonderfully made reduction drive which, through a maze of belts and pulleys, tamed the high speed of the motor down to a slow and steady surge of power which to this day has never faltered. Don and Carolyn Hurless, my nearest neighbor AAVSOers then took over the interior decorating; painting the entire telescope and pier, the observing ladder, and the transit room walls. They gave this bleak, drab room a homey, lived-in look by bringing me curtains for the windows and a rug for the floor. One corner of this room can be partitioned off with sliding drapes and has a large electric heater for quick winter warm-ups.

An undertaking such as this is never completely finished, for improvements, refinements and changes are always suggesting themselves, and this is precisely as it should be, for perfection is a stalemate. The observatory now was fully usable. It had all the necessities for serious observing—even some of the luxuries—and it performed smoothly and with youthful vigor. From the very beginning the project had been an adventure in friendship. From that first phone call in the night to the last leisurely dab of paint it had been an epic of generosity and helpfulness.

The transplanting of the observatory was one more chapter in a story that had begun nearly a century before, for it was early in 1868, just three years after the close of the Civil War, that Professor J. M. Van Vleck commissioned the firm of Alvan Clark & Sons, of Cambridgeport, Massachusetts to build a 12-inch telescope and equatorial mounting for Wesleyan University, of Middletown, Connecticut. The telescope was completed by late September and was tested by Professor Winlock of Harvard and by Professor Watson, the eminent discoverer of minor planets, who approved it with the solemn pronouncement that it "performed admirably."

At Wesleyan the telescope was housed in an observatory that was topped by a revolving cylinder rather than a conventional dome and this was located on top of a three-story dormitory building. Here it was used until 1914, when the new Van Vleck Observatory was built. A 20-inch Clark refractor had been ordered for this new observatory but World War I was then in progress and since all raw optical glass was produced in France, the finished objective was not received until 1922. During these years of waiting the 12-inch objective had been used in the 20-inch tube, being placed at its proper focal distance of 15 feet, 7 inches from the eyepiece. Here it was used visually, and also photographically with the addition of a Number 12 filter.

Upon receipt of the 20-inch glass, the 12-inch instrument and equipment was offered for sale and was purchased by Miami University and there a new observatory, known as McFarland Observatory was built for it. Professor William Anderson, who had retired from active teaching several years prior to our first visit to Miami, related to me how he had made the trip to Connecticut to inspect the equipment and, after approving it, had returned to Miami, bringing with him the 12-inch objective securely packed in his suitcase. Nearly forty years later the 12-inch was to make still another such journey to a new frontier—this time wrapped in a blanket on the rear seat of my car. Dr. Anderson also divulged that he had been responsible for the tiny chip in the edge of the objective. It had been done, he said, by getting one of the adjusting screws too tight while fitting the lenses in their cell. He told how he had then written to the J. W. Fecker Company of Pittsburgh to find out the cost of replacing the chipped crown element with a new one. When Fecker quoted a price of five thousand dollars for just one of the two elements that together make the complete objective he decided to pursue the matter no further. Actually the chipped edge does not affect the image quality of the lens in the slightest degree.

Since finally settling in its new home the 12-inch has done its level best to show off its accomplishments and as yet I have not ceased to marvel at the wonders it reveals. Star clusters such as M 13 in Hercules and M 11 in Scutum are gorgeous quite beyond description, and these are only two among a host of these faraway star-cities whose sparkling street lights seem to wind and twist about until they fade out in the distance. A favorite of mine is known as NGC 4565, the edge-on galaxy in Coma Berenices. Still another is the weird and ghostly Ring Nebula in Lyra with its faint and difficult hot blue star in the center of the ring. M 42, the Great Nebula in Orion, is breathtaking in its sharply defined bright and dark nebular cloud forms. All of these celestial show-offs I had watched hundreds of times before in my other telescopes, but with the 12-inch, everything that before had been vague and elusive was now sharp and clear. It was pleasant to make their acquaintance all over again.

Shortly after the new telescope was in operation an unusual event occurred that well demonstrated the advantage we had gained by the four-fold increase in light-gathering power. A supernova had suddenly flared up in one of the many faint spiral galaxies in the constellation Virgo. Each year several such outbursts show up on photographic plates but only rarely does one become bright enough to be seen as an individual star. This one became as bright as the twelth magnitude and Don, Carolyn, and I watched it for more than a month, making many estimates of its changing magnitude, until it finally faded out and was lost in the background haze of the nebula. For each of us it was immeasurably the most distant star that we had ever seen and it was our first glimpse of a star in a galaxy other than our own. All this would have been quite beyond the powers of my smaller scopes.

It is in the seeing of faint stars that I have derived the greatest benefit from the use of the new telescope. With the 6-inch it was only on nights of the most exceptional transparency that I could glimpse a star below the fourteenth magnitude. With the 12-inch, under the same conditions, I can see slightly below the sixteenth. While this may seem a relatively small gain it must be remembered that each decreasing magnitude is two and one-half times fainter than the preceding one, and that to gain another two magnitudes, or eighteenth magnitude, would tax the visual capacity of even the 40-inch glass at Yerkes. The two magnitudes—from fourteenth to sixteenth—are vital ones to me in variable star observing, for the minima of many of the most interesting variables that I observe lie within the range of these two magnitudes. Formerly I would watch these stars as they slowly faded until, at about fourteenth magnitude, there would eventually come the night when they no longer could be seen. Then weeks or even months would pass before a tiny point of light would once again announce their reappearance. Now, happily, all this is changed, for only two or three of all the stars I watch ever sink so low that I cannot follow them in all their darksome doings.

The new observatory has now been in use sufficiently long for everything to be completely routine. I find that my observing time is about equally divided between the two telescopes and they have, thus far, seemed quite compatible. It is hoped that, as time goes by, additional observers and a diversity of programs will keep the 12-inch still more fully occupied. Already, Oliver

Lundgren, my photographer friend, frequently uses it to take exquisite pictures of the planets and the moon.The motions of the moon still perplex him just a bit but, after all, they even gave the great Newton his only headaches. Don and Carolyn, the variable-star observers, come over at regular intervals and bring with them each time a list of the stars that have dropped below the range of their own instruments and the ever-obliging 12-inch never fails to bring them back in view. These two faithful friends also most efficiently take charge whenever there is a scheduled visitor's night and for this assistance I am truly grateful.

It had taken the 12-inch nearly a century to travel from Cambridgeport to Delphos. Many quite unrelated happenings along the way had helped to bring it to its present hilltop home. Other whims of fate will touch it in the years to come. May its course be always guided by the stars.

A short time after the new observatory had been completed I was returning home after dark from a lecture in a nearby town. At the crossroads where the highway spans the river I turned off on the old country road and drove the quarter mile north with a sudden desire to see what the home farm still looked like late at night. A couple of minutes later I stopped the car on the deserted road in front of the house. I turned off the motor and the car lights and then just sat there, for a time, utterly bewildered by what I saw. Night no longer came to the farm!

Not two hundred yards from the spot where my old observatory once stood a powerful light atop a high pole flooded the surrounding acres with a bluish glare. I got out of the car and looked around. From where I stood I could see five other lights on other farms all spilling out their garish glow. Here at my pasture observatory, during the years when it was the center of my little universe, this midnight sky was seldom shattered by a single ray of man-made light. Today, as evening falls, a sinuous constellation of farm floodlights, like some incandescent Hydra, wraps its coils about the skyline, and glows with baleful eyes throughout the night.

I recalled that years ago I sometimes drove past these same

farm homes late at night. To me each one seemed like a tiny village with its house, its barn, and all its odd array of smaller buildings. But whether I saw it by starlight or underneath the moon, it always impressed me how gently and how peacefully each little village slept.

The moon and the stars no longer come to the farm. The farmer has exchanged his birthright in them for the wattage of his all-night sun. His children will never know the blessed dark of night.

I got back in my car and drove on home, disturbed and saddened by the change that I had seen. I was thankful, though, that I now watched the stars in my peaceful city skies, where my telescopes were safe from the bright lights of the farm.

27 Comet Carvings

More than fifty years have passed since I stood out in the front yard at the farm—first on the snowy ground of January and later in the lush dew-covered grass of May—and watched in silent wide-eyed awe the mighty sweep of those two ghostly visitors of 1910.

Ever since that time I have had a deep concern for comets. It was as though those two forerunners had dispensed some subtle aura that I unknowingly absorbed; some potent emanation which for a time lay dormant until the media was ready, then flared into a lifelong urge which never has abated.

I am quite certain that as I watched those filmy phantoms I had no premonition of the part their kindred comets were to play throughout my life nor did I, at that tender age of ten, invoke a solemn vow that somehow, sometime, I too would find a comet. I did not even know that there were other comets to be found! I only knew that they were something that came out of the depths of space, crept like wraiths across the sky, and then were gone. Something so mysterious and rare that no one, not even my own parents, could explain just what they were.

During all my early teens no other comets came my way. This is not surprising for it would have taken something extra special in the sky for it to be brought to my attention for in those years our only news of any outside happenings came to us on Wednesdays and on Saturdays in the four-sheet Delphos Courant. Even when, at sixteen, I had picked my way to a strawberry spyglass and started out to thoroughly explore the sky I still had no way of knowing where to point that scope to catch a glimpse of one of these strange and fleeting visitors. Not until my winter walk to town which brought the mail to our snowbound farm did I get the slightest clue of where to look for comets in the sky. Included in the bundle of mail along with my first variable-star charts, three letters from my brother Kenneth in France, two *Saturday Evening Posts* and a *Farm Journal* was my first copy of the magazine, *Popular Astronomy.* In it I found, to my great delight, that each issue had a section that told of all new discoveries and also gave the sky locations in which any current comets could be found. Unfortunately, for many months thereafter no wanderers from space came within the range of my little spyglass and it never got to see a single one before it was retired from active duty.

However, the 4-inch came to me from Harvard early in December and soon I had it mounted in the pasture on its white-ash pier. In the meantime, positions of a periodic comet had been published and these I carefully plotted in my Upton's atlas. The object then was crossing Gemini, moving northward toward

Auriga and when darkness settled on the first clear night I pointed the new 4-inch to the east just above the treetops and started slowly sweeping back and forth along my plotted line. Five minutes later—just above the twin stars Castor and Pollux—a little misty blob of light with an even fainter offshoot of a tail appeared against the dark background of sky and, with the same exultant feeling that came to me with the spotting of R Leonis in my spyglass, I knew that this first tracking down a comet with a scope was something I would long remember.

This periodic visitor was already in retreat and never came near naked-eye visibility. Compared to the great ones I had seen in 1910 it was really a sorry spectacle but it was the first one I had seen in nine long years and, fascinated, I watched the eerie traveler for many nights as it moved among the stars. It gave me a most satisfying feeling to reflect that from a maze of figures in a magazine I could point my telescope to a comet in the sky.

Late in 1922 came the 6-inch telescope from Princeton and with it came the firm resolve that I would find a comet of my own. Now I began regularly sweeping the sky on every sparkling moonless night. This was a slow and careful search in which I thoroughly surveyed the entire sky, sketching the location of each suspicious object that I saw, learning to recognize by sight the distant nebulae and clusters, the hazy stars, and the little wisps and blobs of light that so closely mimicked comets. All this was time and effort well expended for, though I little knew it then, I was starting on a quest that would continue for more than forty years.

I already have recounted all my thoughts and all my doings on that never-to-be-forgotten night of Friday the thirteenth in 1925, when my own first comet swam into my field of view. Columbus at San Salvador could have felt no greater thrill. At the outset of this search my original desire had been to find a comet of my own. But a single taste, if it be a pleasant one, seldom satisfies and so the search went on and on. Three more comets—in 1930, 1932, and 1933—were found and though none of them became bright enough to be seen without a telescope, yet each one brought to me the thrilling shock of its first sighting, the hurried sketch to verify its motion, the dash to town to start my message on its way, then the excitement of following the stranger's nightly crawl across the sky, and—after three well-spaced sightings—the making of my own rough prediction of its future course and brightness, and finally came the glow of satisfaction as I carved the year of the discovery in the wooden tube of the telescope.

The night of May 14, 1936, was warm and soft and all the stars were out. It was a night just like that mid-May night some twenty years before when I, from our front yard, first looked up at the stars and realized I didn't know a single one of them. Out at the observatory I removed the shutter from the dome, turned on my ruby light, and for an hour and a half was busy estimating and recording variable stars. Then, to finish off the evening, I turned my comet-seeker toward the north and began a crisscross searching of the region near the pole. Two minutes later the built-in circuit between my eye and arm was broken and my motion stopped. This involuntary reaction is quite often triggered by even a pair of faint stars so closely placed that they present a fuzzy image so I was not excited by this routine ceremony. I brought the blur into the center of the field to have a closer look. It still was just a blur so I shifted to a higher power eyepiece. Even with one-hundred power it refused to separate into a pair of stars and now I felt the tension mounting for I was sure that in my file of comet masqueraders I had listed none in this locality. Out came my sketch pad and pencil and soon I had the object down on paper all lined up in its relation to its nearby neighbor stars.

Then came a period of watchful waiting. Five minutes later I looked at the field and could see no change. Ten minutes more and the thing still seemed frozen in the sky. I often had detected comet motion in less than fifteen minutes so this must just be a faint nebula that I had somehow overlooked in all my former sweeping through this region and, considering that this object was a faint tenth magnitude, I knew that this was possible. I went outside and strolled around a while. In the darkness I could just make out the sleeping shapes of three cows who had settled for the night not far away. I wandered on up to the pump that stood between the house and barn and got a drink. It now was nearly midnight and, returning back along the path, I faintly heard the striking of the clock in town. It brought to mind how comet-hunter Messier, while watching from his home, could hear each hour the clang from more than forty churches.

I went inside to have my final look before I closed up for the night. During my midnight stroll the eastward spinning earth had moved my telescope and now it took a little searching before I found the field that I had left there. But when I did the metal dome above me reverberated with a loud ecstatic "Whoopee"—for that tiny blob of light had moved!

Once again I drove to town and climbed the tower steps two at a time to send my telegrams to Harvard and to Van B. at Yerkes. But this time I didn't drive back home past Dottie's house and give our five short coded toots to tell her of my passing. Dottie didn't live there any more. This time when I returned she was waiting up for me in our own home by the river.

It was at once apparent from the comet's faintness and extreme slow motion that it was very far away when found. More than a week elapsed before it crept out of the original field that I had sketched but all during that June it gradually picked up both brightness and apparent speed. On July first it could be seen without a telescope and throughout that month it cut quite a noble figure as it moved southward toward its closest rendezvous with earth, while each night visitors from far and near paraded up our pasture path and Dad was knee-deep in delight. The nearest approach came in early August when the comet was a mere 15 million miles away. Bright moonlight at the time diminished the full splendor of the comet's tail but the head was a glowing body a full two-thirds the apparent size of the moon.

It had been the brightest comet I had seen since 1910 and I was highly pleased with its performance. One astronomer figured out its orbit to be a long ellipse which would bring it back again in about 450 years. On the night of August 22, I took my last look at the comet and said a fond farewell for on the following day it would sink below the south horizon and be lost to sight. But out on my stubby tube of dark mahogany I had already carved the comet's signature, 1936-A, in figures extra sharp and clear. I hoped that they would still be there when the comet comes back to earth again in 2386.

Another comet, a rather faint one, was located in 1937 and still another two years later. This was the last of the seven to be picked up in the pasture for that was the year we moved to town and the comet-seeker was installed in the merry-go-round observatory. While spinning in this structure five additional comets have been spotted though only one of these reached naked-eye brightness.

In recent years the professional astronomers with their big wide-angle cameras have made great inroads into this field which has always been so closely identified with the amateur observer. In a statistical study of all comets appearing during the twelve-year period from 1948 to 1960 the British observer, M. J. Hendrie, found that amateur observers had made twelve discoveries while the professionals had accounted for fifty-two. Finding comets captive on glass plates must lack the tense excitement which the sweeper of the skies experiences each time he sights a stranger in his scope, but it serves a useful purpose for it captures many faint ones that otherwise would get away.

In spite of this increasing competition there always will be comets for the amateur to seek and, in some facets of this work, he still has an advantage. In a given time he can cover far more sky than can the camera, he can know within half an hour the true nature of a suspected object and he can search much closer to the sun in regions which would fog a photographic plate.

Over the years I have devoted many, many hours to the search for these always welcome visitors from space. They have been leisurely, pleasant hours, and usually indulged in only after my nightly program of variable star observing has been completed, thus, no matter how futile is the hunting, I never have the feeling that my whole night's work has been in vain. It is an ideal leisure-time pursuit. I have repeatedly noticed that its methodical procedures require little if any mental effort or concentration. In time one becomes a veritable automaton. Many a time my physical self has been busily hunting comets, my eye glued to the telescope, my hands endlessly guiding the instrument back and forth across the strip of sky, while my mind, in complete detachment, may be pondering some such weighty subject as the planting of my early spring garden. Then, suddenly, my eye sees something a little different in the star field, my hands automatically stop and my physical being, like a well-trained bird dog, freezes while my slowly grinding mind makes the transition from carrots back to comets.

Time has not lessened the age-old allure of the comets. In some ways their mystery has only deepened with the years. At each return a comet brings with it the questions which were asked when it was here before, and as it rounds the sun and backs away toward the long, slow night of its aphelion it leaves behind with us those questions, still unanswered.

To hunt a speck of moving haze may seem a strange pursuit, but even though we fail the search is still rewarding, for in no better way can we come face to face, night after night, with such a wealth of riches as old Croesus never dreamed of.

To me, comet-seeking is a magic carpet that can take me on nostalgic flights into the past. On the farm, in my early years of comet hunting, few sounds of civilization ever came to me inside my little dome. Even now, when I am hunting late at night, there is little to remind me of what century it is. In the dark silence of the dome two hundred years can disappear in just the twinkling of a thought and I am standing beside destitute, old Charles Messier as he knocks on the door of his friend Lalande and borrows a little oil for his midnight lamp. On another night I may, in spirit, be with Dr. Olbers of Bremen, as he turns from stethoscope to telescope and begins his nightly search for asteroids and comets.

But all too often, just when I have conjured up the proper bygone shade and with it have escaped into the tranquil skies of yesteryear, somewhere up the street an auto horn will give a raucous toot and recall me to the present.

28 Observations

OLD TELESCOPES NEVER DIE, THEY ARE JUST LAID AWAY. THERE IS little about a telescope that can deteriorate and the lens, the vital organ, even though a century old, can still have all the fire and sparkle of its youth. Sometimes, however, for one reason or another, telescopes do become dormant or go into a state of suspended animation, and at such times, as they slowly gather the dust of the passing years, it seems that they must wonder just what has become of the hands which once so eagerly pointed them to the skies.

In this same vein I too have often wondered about all those myriads who, with many an expression of utter amazement and delight, have watched the rings of Saturn, the craters of the moon or the incredible Orion Nebula through these same instruments of mine. Both of these telescopes, even before they came to me, had already served long careers before the public eye. One is a Princeton graduate, the other an alumnus of both Wesleyan and Miami Universities, and each has spent a lifetime in showing-off the skies. So, it would seem quite logical that, out of all these thousands who, at some time, have peered into these telescopes, there must have been at least a few who, in some way, have had the course of their lives changed or affected by the impact of the wonders that they saw, even as I once brushed against some stars which left their lifetime mark on me.

These telescopes of mine are not just bits of property. I like to think of them as gifts in trust and myself as their administrator; one who is grateful that it has fallen to his lot to open up their eyes and let them see the stars again. They now seem quite rejuvenated and, with youthful zest, they look ahead to long and useful lives. But I am not misled by their appearance. I know something of their past and, in spite of all their sprightly mien, I still regard them with that feeling of deep reverence that one accords a patriarch, for they have witnessed much of history. Especially is this true of the 12-inch for I am much more certain of its genealogy.

So many changes have occurred since it left the deft fingers of the Clarks and first looked out upon the world of 1868. But, though now almost a hundred years have passed, it is only here on earth that much change can be detected. Farther out in space our neighbor, moon, has shyly edged a scant three feet further from the earth. Halley's Comet made one brief appearance, swung its tenuous tail around the sun, and then was gone again. Pluto, long suspected, came from hiding, and Neptune made but little more than half a circle round the sun. Deep in space Barnard's runaway star moved among the other stars one-half the apparent diameter of the moon. Nearly all the other stars seemed frozen in the sky, though now and then a nova spent its substance in a futile flash, then settled down and sulked.

The cosmologists were vastly busy during all these years in scrapping old ideas and inventing brand new models. The 12-inch had been born in the long reign of the nebular hypothesis, it grew up in a rain of planetesimals and it matured during the leisurely eons of continuous creation and the bang of the primordial atom. Its golden years will doubtless be spent amid favorite phantoms equally interesting and equally ephemeral. Man is a most audacious artist. Only yesterday he was drawing pictures of wild beasts on the dimly lighted walls of his cave. Today he would picture all the universe and how it came to be!

Of all the signs of changing times which my telescope has seen, the ones it views with ominous concern are neither on the earth nor in the depths of space. So often, of late years, strange lights pass across the star-field I am watching through the telescope. Some of these lights are bright, still others are so faint that they have halfway crossed the field before I notice them. Nearly all of them have been moving in an easterly direction and for all that I have seen there comes a moment when suddenly they fade and disappear, and by this I know the thing I have been watching was lighted by the sun and now is lost within the long black shadow of the earth. All too well I realize that I have been witness to mankind's latest pollution in the name of progress—the contamination of the skies.

Already, in only the eighth year of the Space Age, the sky is littered with a jumble of jettisoned nose cones, carriers, and drop-off stages from the multitude of spacecraft placed in orbit round the earth. Unlike the graveyards of the Automobile Age, whose slowly oxidizing carcasses can at least be zoned, these cerements of space will circle uncontrolled across the skies until, in time, the pressure of light and the slight resistance of the upper atmosphere will slow them down until at last, as man-made meteors, they will make their final fiery plunge, trailing behind them a wake of ashes to continue the contamination.

In these strange lights that cross the sky my two scopes see a gloomy portent, a distant early warning of the nights to come. Forty years ago, on the top of Mt. Wilson, the world's largest telescope could look down and see the gathering lights below. Today the approach is from above as well.

So much that man touches he destroys. Less than five centuries ago three tiny ships dropped anchor off the coast of a little island and there began the greedy pillage of the Americas. Today man is reaching for the moon with those same eager hands of conquest. Already we have been informed that the moon is a challenge which we, as a nation, must meet.

Time after time we have been told the story of the noted mountain climber who was asked why he felt impelled to scale a particularly difficult peak. His reply was simply: "Because it is there." A less dramatic but more forthright answer would have been: "Because my ego demands it." The conquest of the moon, as planned at present, is simply a higher mountain with, at its summit, the same dubious prize—prestige.

I know that someday man will reach the moon but I sincerely hope this will not happen for a long, long time. He has a lot of growing up to do before he will be ready for the moon. When he finally does set his sails for a journey into space, may it be a voyage of the Beagle, not the Graf Spee. If man must meet a challenge he can find one here on earth. If he must conquer something let it be himself.

The moon and I have been firm friends for all these many years. On a thousand nights I have turned my optic tube to her and she has gaily entered. From my little observatory, in contour-cushioned comfort, I have explored her rocks and rills, her plains and craters, and I have watched a thousand times the play of light and shadow on her face. In our western wanderings we once stood spellbound as she rose, incredibly large, above the sands of Alamogordo, and on another night we saw her slip into a midnight bath in the Pacific, her face half hidden in her shadow.

Tonight, as lustful eyes are turned in her direction, I see again the gently smiling face of Lady Moon, my early love of Second Reader days and once again I hear the repetition of her soft reply—"All that love me, all that love me."

The light has all but faded from the sky as I start along the hilltop path to the observatories. From somewhere in the shadows my big collie, Canis Major, noiselessly appears and trots along ahead. We round the big lilac and I pause for just a moment at the small observatory to throw back its lid before going on to open up the larger one. Spring and summer both have gone and there is more than just a hint of autumn in the air. The tempo of the cricket's chirp is beating slower night by night and surely frost can not be far away. Far to the south lone Fomalhaut is marking out a trail for winter stars to follow and through the eastward-opened dome I see the Pleiades.

Even as I watch, the dome above me fades away and now its opened shutter is a darkened kitchen window through which I gaze in childish wonder at seven little stars that sparkle in a long-gone autumn sky.

Index